For H.

This is a work of fiction. Names, characters, places, and incidents either are the product of the author's imagination or are used fictitiously. Any resemblance to actual persons, living or dead, events, or locales is entirely coincidental.

Copyright © 2020 David Monteith

Illustrations copyright © 2020 Nikki Dawes

Cover and layout by Nikki Dawes

All rights reserved. No part of this book may be reproduced or used in any manner without written permission of the copyright owner except for the use of quotations in a book review.

ISBN 978-1-73314-77-0-5

Table of Contents

-	**Prologue**
1	**Chapter I**
9	**Chapter II**
13	**Chapter III**
21	**Chapter IV**
27	**Chapter V**
33	**Chapter VI**
37	**Chapter VII**
45	**Chapter VIII**
50	**Glossary**
55	**About the Creators**

E ven butterflies have legends. Stories they share while gathered together. Tales they tell to inspire and explain. This is among the oldest of the butterfly legends. Like most tales of butterflies, it is a story of change. It tells how things that are came to be.

But first you should know how things were...

Prologue

Long, long ago, before there were so many people, in so many houses, with so many cars, there was nature. Birds and butterflies filled the skies like a kaleidoscope of colors and wings. Flowers and tall grasses covered the plains. Animals roamed freely. In many ways, life back then was idyllic – idyllic but not without violence. Even then, lions hunted gazelles and snakes hunted mice, but far fewer animals preyed on one another in those days. With so much vegetation, they had no need to.

And in those early days nothing hunted butterflies.

Butterflies looked then much like they do now with one distinct difference – they rarely flew. Being lighter and much less strong than birds, any significant wind would blow them off course, so butterflies of old often hopped from flower to flower. They only used their wings to give themselves a tiny boost off petals or the ground. Sometimes they would wait for a light breeze then hold their wings stiffly out to the side and glide like colorful paper airplanes to nearby flowers or bushes. At the first sign of a strong wind, they would fold their wings flat behind them and wait for it to pass. They lived a slow, leisurely life.

Until the Dead Season.

No one knows why, but summer that year burned hotter and longer than it ever had before. Streams dried. Flowers died. Animals and insects migrated farther and farther north seeking relief from the heat, but there was none to be had. Food became scarce and things changed. Birds who once joyfully shared the skies with butterflies now saw them as easy prey. Desperate from hunger, the crows were the first to turn, then the starlings, the jays, and the kingbirds. In flocks they hunted and fed.

In a few short months, millions of butterflies were reduced to thousands scattered across the land. Small groups of butterflies led by one or two elders hid where they could and tried to plan for survival. This is the story of one of those butterflies in one of those groups and it begins with...

"Ned!"

The terse whisper called Ned's attention back to Stern, the leader of his group.

Ned's antennae drooped flat. "I'm sorry, Mister Stern, sir. It's just...that cloud in the distance. It's so still and peaceful. I was thinking how nice it would be if we could all be at peace like that cloud."

A flutter of nervous giggling bubbled through the group of young butterflies. Stern turned his head to follow Ned's gaze.

"That's snow," Stern said flatly, "on a mountain."

More snickering erupted.

Ned knew he didn't see the world like the other butterflies. He didn't understand why everything appeared blurry unless he was nearly next to it, but it did. It always had, so Ned chose not to let it worry

him. He liked the world as he saw it.

Gram, the group's other elder, intervened before Stern could lose his patience.

"That's a lovely wish for us all," she said. "And one that we can make real if we pay attention and listen carefully to Stern."

Following Gram's lead, Stern began the lessons again. "The shortest distance between two flowers is..."

"...a straight path," the younger butterflies responded in unison.

"Straight paths come from..."

"...stiff wings," they replied.

"You will be afraid," he said, looking at each one of them. "It's natural, considering what we're experiencing. When you feel scared repeat these mantras. They keep the fear from taking control." A caw in the distance sent a shiver through the group.

Stern emphasized his point by continuing, "Looking up..." he prompted.

"...slows you down," they answered.

Gram's mind wandered past the daily lessons. She looked at the group of young butterflies and wondered how many they'd lose today. No matter how well they followed Stern's directions, there was no escaping their lack of speed. Butterflies always had been and always would be slow.

The cold made matters worse. They'd flown far north searching for food and relief from the heat. She'd hoped – they'd all hoped – the cooler temperatures of fall would last as long as the long, hot summer.

Instead, fall fell quickly. The transition from hot to cold happened almost overnight. Geese riding the north wind announced the coming of winter.

As if he was reading her mind, Stern reminded the group, "We can escape the birds, but we can't escape the cold. We must keep flying south. Don't give up. There are others out there. Flowers and friends are at the end of our journey." He hoped he sounded more confident than he felt.

Stern's lessons concluded and he prepared the butterflies for the day's journey. He pointed out the various bushes and trees they'd use for cover. The cold winds were no friend to the butterflies, but at least they'd brought a change of color to the trees. The red-gold leaves were similar enough to the coloring of their wings to offer a bit of camouflage.

"Follow Gram's lead," Stern commanded.

Quiet and reassuring, Gram always flew at the front of the group while Stern followed from behind. It allowed him to call out course corrections and keep an eye on those butterflies whose fears might lead them astray. From the back, his encouragements and reminders were carried forward on the wind when he saw the young ones falter. It also meant he witnessed every butterfly that had been plucked from the sky.

II

After another day's travel, they stopped for the night at the base of a hill. A lone tree stood at the top of the hill, but Stern chose to rest for the night in a small group of bushes that offered more protection.

"In the morning we'll fly up to that tree and plot our course for the day," he instructed. Had they not been so tired, the young butterflies might have noticed the look in his eyes as he glanced at Gram. Within minutes they were all asleep. The wind was still and the night was quiet.

"Only three," Gram said, looking over the group of resting butterflies. "I never thought losing three butterflies would feel like success, but I'm as grateful as I am sad tonight."

She ended every day by naming something she was thankful for, some piece of brightness she could hold onto. She did this aloud, sometimes with the younger butterflies, sometimes feigning talking to herself, but always within earshot of Stern.

He'd never admit it, but she could see the journey pulling something inside him further and further down every day.

Stern had difficulty seeing anything but failure. He took every loss personally. He knew blaming the birds or the north wind wouldn't help, but he couldn't keep those thoughts from entering his mind. Uttering them aloud would only give them more weight, he thought, so he kept them inside. Like the mantras he taught the young ones, he told himself over and over again to focus on the path ahead.

"Do you remember what's on the other side of this hill?" Stern asked.

Gram nodded. They were returning south following the same path they'd come.

"You'll think of something," she said.

"No," he sighed, "I won't."

Unsure of what to say, Gram moved closer to him, offering silent support. They sat there a while, side by side in the dark.

Stern forced himself out of his negative thoughts with a wry chuckle, "A cloud, eh? How a basically blind butterfly has managed to stay alive through this is a mystery to me."

"That he still sees some beauty in the journey is what amazes me," said Gram.

They watched over the group in silence for a bit longer before falling asleep themselves.

III

They went to sleep thinking about Ned and the journey ahead, but woke up thinking about birds.

Dozens and dozens of birds *cawed* loudly in the tree at the top of the hill. Stern's antennae moved slowly, commanding stillness. He could feel the terror in the young butterflies as their eyes snapped open.

"Wings folded flat and tight..." he whispered, knowing each of them in their minds was replying, "...keep us all out of sight."

If the birds had seen them, they'd already be dead. *Just keep still*, he thought.

The cacophony of cries seemed to last forever.

Sharp, cold gusts of wind whisked through the bushes reminding them they'd have to move eventually.

To their credit, the young butterflies, even Ned, held the stillness until, without warning, the birds exploded as one out of the tree.

Stern watched as the starlings moved in one giant murmuration, driven by their own southward urging. Slowly, he spread his wings, indicating to the others was safe for them to move.

"Well done," he said as Gram moved among them comforting each one. "Well done."

Gram led the solemn group to the lowest branches of the tree. A few dying leaves clung to the branches, tiny warning signs that winter was near.

"As difficult as our journey has been so far, today will be our most challenging day," Stern began.

He didn't need to explain why. Every butterfly saw the empty field stretching out ahead of them. There would be no cover from trees or bushes during this flight, and walking would leave them exposed for far too long.

At the other side of a long, downhill slope lay the edge of a thick forest, but it seemed impossibly far away. What would have once been a leisurely glide over a flower-covered field was now the most dangerous scenario they could imagine.

"We must fly straight and true to have a chance against the birds. It's our only hope, because we can't survive the cold. Let's review your mantras," he said.

A small flock of birds appeared in the distance. Normally willing to follow Stern and Gram without hesitation, the younger butterflies' fears erupted. Questions burst over one another as they all began speaking at once.

"How…," "What if…," "What should…," "Do we have to…," "Can't we…" They wouldn't or couldn't be calmed.

Four shadows crisscrossed the field, hinting at the birds above. The younger butterflies became more and more resolute in their refusal to fly.

Ned stared out across the open field. He was as frightened as the others, but he didn't see how adding his voice to the mix would ease the chaos and confusion. The forest's edge was a vague blur to him. He knew by their shapes they were trees, but he could hardly tell where one ended and the next began. *Safety,* he thought, *is so dark and far away.*

The shadows slid like snakes across the field. He knew he shouldn't, but Ned looked up. He wanted to find a quiet cloud, something peaceful to rest his mind on. He didn't see clouds though, and he didn't see birds.

Instead he saw a lone butterfly perched high in the branches above. Knowing the danger of being in easy view of the birds, Ned called out.

"Come down," he whispered.

Maybe she doesn't realize where she is, he thought. *Or maybe she's given up. Maybe she just doesn't care anymore.* Without thinking about it, Ned climbed toward her. He whispered again. "It's not safe up there. Come down!" A gust of wind took the words just as they left his mouth.

She started to move and Ned froze. Instead of walking or flying like other butterflies, this butterfly swayed with a gracefulness he'd never seen. Her coloring and her movement reminded him of a slow flame. Beautiful and carefree, she hung lightly from the branch, gliding between quick, precise twists and languid turning without pattern or predictability. Ned was mesmerized. The fearful chattering of the other butterflies faded away.

He started to climb toward her again. Torn between wanting her to continue and getting her back down to safety, he couldn't take his eyes off her dance. She was so free and unafraid.

A dark blur appeared overhead.

"Come down!" he whispered again. She swayed, and turned, and twisted and then-

She leapt!

IV

"No!" yelled Ned.

His world slowed for a moment as he watched her float away on the breeze. He couldn't believe she'd left the safety of the tree. Fear and confusion and sadness swirled like a cold wind through Ned. He tried to look away, certain the birds would snatch her from the sky, but her gracefulness caught his eye again.

She wasn't gliding stiff-winged across the field. She bounced and bobbed through the air just as she had on the branch. She was beautiful and she was fading from his sight. In a moment of perfect clarity, Ned knew he couldn't lose her. He spread his wings and he leapt.

"Ned!" He heard Stern shout, but if anything more was said, it was lost in a gust of wind.

Ned wouldn't lose her. Stubbornly, stiffly, he tried flying a straight path to her as she danced her way across the field. *The shortest distance...* he told himself, keeping his eyes focused on her.

Straight paths come from... he grunted and strained to keep his wings stiff. Changing course was usually a matter of a slight shift in the wings for butterflies. *Tiny tilts for large turns*, was another of Stern's mantras.

Small adjustments were usually effortless for Ned, but he'd never had to make so many so quickly. The strong wind punched him back and forth. She, on the other hand, bobbed and twirled so gracefully and so unpredictably. After a few moments into his flight, Ned was exhausted from shifting course so many times. The distance between them grew.

How was she moving so quickly? Ned realized he

could never catch her by gliding. Out of desperation and against every lesson he'd learned, he began flapping wildly. He imagined how erratic he must look to the other butterflies. He was sure he could hear their laughter on the wind.

V

But there was no laughter back in the tree. The others had stopped their arguing once they'd seen Ned leap. Minutes passed as they silently waited for him to be plucked from the sky. Despite their lessons, they couldn't help sneaking glances into the air. The birds circled and readied themselves to dive.

Stern and Gram looked at each other, not knowing why Ned had leapt, but knowing they could do nothing to help him. Stern could neither understand nor deny what he was seeing. Why had Ned leapt? Why was he flapping wildly through the air? Why weren't his wings stiff and straight? And, most importantly, why weren't the birds attacking a lone butterfly on an open field?

Ned might have wondered the same thing if he hadn't been so tired. Putting so much effort into flying was new and exhausting. Muscles he'd rarely used began to burn. Before long he didn't have the energy to think about mantras or about what the other butterflies thought. He hurt too much to look up at the birds or notice their shadows down below. He strained to catch the graceful butterfly as she floated away.

Ned wasn't even halfway across the field when exhaustion overtook him. Discouraged to the point of despair, he knew he had no choice but to land and rest. He tried to tell himself he could continue following her by walking across the field, but inside he knew the distance between them would only grow if he was forced to move that slowly.

As he and every other butterfly had been taught, he flattened his wings behind him. He would drop rapidly, and then, at the last moment, spread his wings to stop his fall and land gently on the ground.

Speeding quickly toward the field below Ned kept his eyes on the butterfly still floating away. All of his effort hadn't been enough to catch her.

Dejected and near the ground, Ned spread his wings like a parachute. And then it happened. Instead of dropping lightly to the earth, Ned found himself scooped up by a strong gust. The wind caught and carried him up and up and up. The rapid, unfamiliar movement scared him. He'd never been swept into such a great current. Uncertain of how to react he held his wings out and was carried along. The wind swirled him up, down and around, faster than he'd ever flown. It was terrifying and strangely effortless.

And then it was gone.

With as much surprise as it had arrived, the current left. Ned was high in the air again. He flapped erratically, desperately reaching, hoping to be carried again. Exhaustion overtook him once more. And once more he folded his wings flat and fell.

Maybe because he'd felt it once already, Ned sensed the approach of another strong current. With nothing to lose he snapped his wings out and whoosh! The wind hurled and swirled him forward and up.

Again the current shifted unexpectedly. Again Ned fell, and again he found a new gust to ride. With each new burst the speed and unpredictability became less terrifying. Flying became less about controlled gliding and more about intuitive falling. He might have enjoyed it if it weren't for his mysterious butterfly. His heart flew higher and faster than any gust of wind when he caught sight of her again.

She was nearing the edge of the forest. A combination of awe, pride, and hope electrified him. If *she* made it then he could. If he made it, he would find her. If she was safe, then he could find her.

Over and over again he fought the temptation to fly directly toward her. He told himself making good speed on the currents was more important than making a straight line to her. *As long as I'm getting loser,* he thought as he allowed himself to be carried chaotically along.

VI

Ned was over halfway across the field and the birds still hadn't attacked. Stern didn't understand why, but instinctively he knew it wasn't the time for wondering. He looked at Gram and, without a word, they decided.

"Follow me!" Gram leapt into the air. She flapped and tilted and bounced, doing her best to mimic Ned's motion.

"Do as she does," commanded Stern. "No more stiff and straight! Ned's making it. Gram's making it. You'll make it!"

He pushed the younger butterflies one by one into light before questions and doubts paralyzed them.

"Don't think!" he called out from behind them. "Just fly!"

And they did. They flapped desperately until, one by one, each of them found the rhythm of the currents as they made their way across the field. From the back Stern could see the moment when they shifted from wild flapping to an energetic, but calm, riding of the wind. Had they not been afraid of the birds above they might have been amazed at how easily they adjusted to their flopping, bouncing flight. They might have even enjoyed how natural it felt.

Natural to everyone but Stern. He was the last to leave the tree. It was a struggle to break the habits he'd been teaching for so long. He found himself glancing up at the birds. He tried to follow Gram's lead, but every tilt of his wings took more mental than physical effort. Every minor change in direction felt enormous to him. More than ever before, he was afraid.

Shadows moved across the ground, reminding him

of the birds circling above. He tried to keep his voice calm as he called out "Looking up…"

If a reply came from any of the younger butterflies, he didn't hear it. He saw them getting farther and farther ahead of him. He allowed himself to stiffen his wings and glide. Just for a moment, he told himself.

Gliding allowed him to gather himself and survey the situation. Ned had landed. So had Gram and several of the others. The rest were over halfway across the field. Relief and joy spread through him from wingtip to wingtip.

In that instance, a new thought passed like a shadow over his happiness.

He thought about how easy it would be to glide stiff-winged as far as he could. He thought about letting the birds take him. He thought about giving up. *They're safe*, Stern thought. *That's all that matters.* He looked again and saw those who had landed calling out to those who were still floating and flapping to safety. He saw a shadow moving on the ground below. Wings stuck straight out from his sides, he stared ahead and continued to glide.

It might have been shame that tilted one wing and then the other. Or maybe the sight of Gram gathering the group in safety woke him from his dark trance. Regardless, Stern was embarrassed by his thoughts. Shame turned to stubbornness as he grunted and strained his way through re-learning how to fly. His old mantras faded away as he told himself again and again, *Don't give up. Never give up.*

VII

Exhausted, Stern landed at last. All the other butterflies gathered around him, grateful and energized, chattering on about avoiding the birds and learning how to ride the wind. All except Ned.

Ned stood apart from the group staring into the forest. Gram caught Stern's eye long enough to indicate she had no idea why. She nudged the others deeper into the safety of the forest while listening to them share their versions of the experience.

"Ned?" asked Stern.

"Did you see her?" Ned asked. His eyes never left the forest. "She was the most graceful butterfly I've ever seen and I've lost her."

Stern glanced around, counting the butterflies. They were all present and accounted for. He looked at Gram who shook her head in confusion.

"Maybe she kept flying. Should we try to catch up with her?" Ned asked.

A leaf, brown and papery, floated by on the wind. Stern saw hope and disappointment flash across Ned's face as it caught his attention. Orange-red leaves on the forest floor shuffled and rustled restlessly. In that moment Stern realized what led Ned across the field, what inspired his flight.

"She was so beautiful. Did you see the way she flew? Where did she go? Do you think she's safe?"

The questions tumbled out of Ned, each so sincere and full of longing that the other butterflies had no heart to mock him. They gathered around, looking to Stern for guidance, wanting to comfort Ned, but not knowing how.

"Leaves," said Stern, a little more gruffly than he intended.

A look of understanding flashed across Gram's face. She started to interrupt, but Stern continued.

"She," and he emphasized this word while giving a firm look to every other butterfly in the group, "She did what beauty in nature always does." His tone softened.

"It inspires awe and wonder. It gives us something to appreciate and aspire to. It moves us. But then it leaves. Sometimes as quickly as a flash of lightning, sometimes as slowly as a cloud or a setting sun. Beauty takes a moment of our breath and grows it into an idea, or a feeling, or an inspiration. But beauty in nature always, always leaves."

Stern paused.

"Your butterfly was true beauty, Ned. It was a brave

thing you did following her across that field. You, and she, led us to safety."

At this Stern's wings spread slowly open. The other butterflies gathered silently around Ned, their wings open.

"Thank you," each of them said quietly. "Thank you."

A cold breeze blew through the group, reminding them the journey wasn't over.

"Winter is still approaching, and we've a long way to go," said Gram softly. She motioned them to find a place to rest. The short days of winter were nearly upon them, and the light had started to fade.

They huddled together for warmth and comfort, but Ned stood apart from the group, staring into the forest. He looked into the darkness long after the day was done. Gram and Stern made certain he knew they were nearby, but left him to his vigil.

Neither of them knew quite what to say. Unspoken questions passed between Gram and Stern in the silent language of old friends.

"Will it break his heart to know she was a leaf?"

"Would it be wrong not to tell him?"

"Will he realize it eventually?"

"Will the younger ones make fun of him?"

A shrug of a wing or a shake of the head told them that neither had the answers. There was no decision that wouldn't break their hearts or leave them wondering if they'd done the right thing.

Stern finally broke the silence of the conversation. "How much longer do you think the birds will be fooled?"

"Long enough," said Gram. "Look at them, all of them. We didn't lose a single one today."

They sat in silence until the darkness was so deep that Ned relented and rejoined the group. The band of butterflies rested.

VIII

The legend doesn't tell us whether Ned was ever told the truth of the leaf he'd followed across the field. We only know they didn't tell him that night, or the next, or the next.

Ned continued to look for his beauty as he and the others flew south like leaves on the wind. Progress was safe but slow as Stern deviated from their original path, stopping at each patch of red-gold leaves that caught Ned's eye. Along the way they found other groups of butterflies huddled and afraid.

Stern and Gram taught each group their new bouncing, flopping flight. When asked how they discovered it, Gram would tell the story of a brave butterfly pursuing a mysterious beauty across a dangerous field. Stern resumed his lessons, omitting some old mantras and including at least one new.

At the end of each day's lesson, Stern reminded each gathering of butterflies, "Find what lifts you up…"

"…and follow it as long as you can."

To this day, butterflies tell the story of Ned the nearsighted butterfly when teaching their children to fly.

The end.

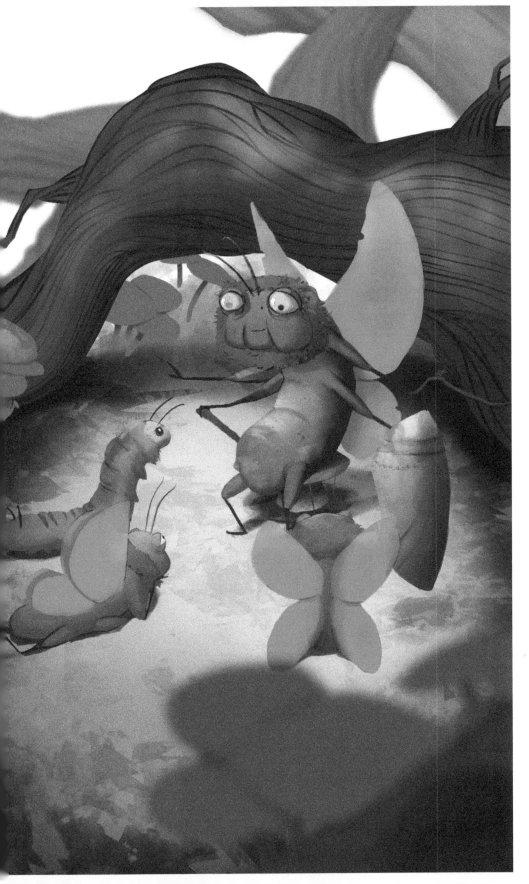

Glossary

Awe (noun) – a strong feeling of fear or respect and also wonder.

Cacophony (noun) – unpleasant loud sounds.

Chaotically (adverb) – in a state of complete confusion or disorder.

Dejected (adjective) – sad because of failure, loss, etc.

Despair (noun) – the feeling of no longer having any hope.

Deviated (verb) – to do something that is different or to be different from what is usual or expected.

Discouraged (adjective) – less determined, hopeful, or confident.

Erratic (adjective) – acting, moving, or changing in ways that are not expected or usual.

feigning (verb) – to pretend to feel or be affected by (something).

idyllic (adjective) – very peaceful, happy, and enjoyable.

intervened (verb) – to become involved in something (such as a conflict) in order to have an influence on what happens.

intuitive (adjective) – a: agreeing with what seems naturally right b: easily and quickly learned or understood.

kaleidoscope (noun) – a changing pattern or scene b: a mixture of many different things.

languid (adjective) – showing or having very little strength, energy, or activity.

Leisurely (adjective) – not hurried; slow and relaxed.

Mantra (noun) – a word or phrase that is repeated often or that expresses someone's basic beliefs.

Mesmerized (verb) – to hold the attention of (someone) entirely; to interest or amaze (someone) so much that nothing else is seen or noticed.

Murmuration (noun) – a flock (of starlings).

Paralyzed (verb) – to make (someone or something) unable to function, act, or move.

Resolute (adjective) – very determined; having or showing a lot of determination.

Scenario (noun) – a description of what could possibly happen.

Solemn (adjective) – sad and serious.

Terse (adjective) – brief and direct in a way that may seem rude or unfriendly.

Trance (noun) – a state in which you are not aware of what is happening around you because you are thinking of something else.

Unison (noun) – If people do something in unison, they do it together at the same time.

Unpredictability (noun) – not always behaving in a way that is expected.

Vague (adjective) – not able to be seen clearly.

Wry (adjective) – showing both amusement and a feeling of being tired, annoyed, etc.

About the creators

David Monteith - Author

David loves playing with words. He participates in National Novel Writing Month every November. He loves flash fiction, poetry, and making earrings out of comic book word bubbles.

dmonteith.com

Nikki Dawes - Illustrator and designer

Nikki enjoys creating illustrations, learning about art, and playing video games and table top role playing games.

nikkidawes.com

CPSIA information can be obtained
at www.ICGtesting.com
Printed in the USA
BVHW020712130220
572027BV00060B/1160

Hwang Tin Nei Jen Jing
The Interior Yellow Court Scriptures

The Internal Scenes of the Spirits of the Organs

Compiled from the writings of:
Madame Wei Hwa Tsun • Pu Ming Chan Shi • Wu Chong Xu

Translated and amended by
Shifu Hwang

Published by:

Shifu Hwang
132 Sage Cove
Bastrop, TX 78602
USA

ISBN 978-0-9851028-2-1

copyright 2015   ©   Shifu Hwang

All right reserved. No part of this publication may be reproduced, stored in a retrieval system, transcribed in any form or by any means, electronic, mechanical, photocopy, recording, or otherwise without the prior written permission of the publisher.

WARNING: When following some of the techniques given i this book, failure to follow the authors instruction may resu in side effects or negative reactions. Therefore, please be sure to follow instructions carefully for all techniques and modalities. If you have any questions about doing any of these techniques safely, please contact a Taoist Master in your area.

DISCLAIMER: The information given in this book is given good faith. However the author and the publishers cannot b held responsible for any error or omission. The publishers will not accept liabilities for any injuries or damages caused to the reader that may result from the readers acting upon or using the content contained in this book.

Acknowledgements

I would like to thank Leslie Mitchell for her help in bringing this book to fruition by organizing, editing , typesetting, designing, and preparing it for print.

Table of Contents

Translator's Foreword 1

Introduction 3
  Building the Foundation 4
  Processing Jin to Become Qi 4
  Process Qi to Feed Shen 6
  Process Shen be United with the Void 8

The Authors and Interpreters 11
  Madame Wei Hwa Tsun 11
  Len Chan 12
  Pu Ming Chan Shi 13
  Wu Chong Xu 13

Hwang Tin Nei Jen Jing (Interior Yellow Court Scripture),
The Scenery of Internal Spirits 15

The Method of Mind Discipline 63

The Secret Phrases for Three Levels Of Nei Dan Training
and Its Interpretation 85
  The First Level 86
  The Middle Level 95
  Use Correct Thought For Distracting Evil Associations 97
  The Secret Phrases of the Upper Level
  Transcendental Method 99
  The Secret Phrases of Returning to the Void 100

Index 105

Translator's Foreword

From the time that I translated the Tao Te Ching to English, in 87, I had a desire to translate more Taoist Scriptures. I wished to read the seeds of Taoist knowledge through the North American ontinent. This book, The Tao Tsan, The Taoist Treasure is the cord of those seeds of knowledge.

Many years elapsed before I could begin the translation of this owledge. My secular life remained unsteady, causing my plans translation to move forward slowly, at best. Nevertheless, I was le to complete the translation of Yin Shi Zi's, Jin Zwo Fa anquil Sitting), in 1993; and Chang San Feng's, Nien Dan Mi Jue ue Dragon, White Tiger), in 1995. Most recently, I completed the nslation of Zao Yi Zen's, Wai Ke Xian Feng Mi Ji (Herbal rgeon), in 2010.

Hwang Tin Nei Jen Jing (The Scenery of Internal Spirits) is a ry important, secret book for Taoist practice. The original author is a female Taoist living in the Chinese Jin Dynasty (284 - 330 D.). Residing in the mountains of Mao Shan, she intensively ltivated her practice of Nei Dan training.

In order to further my understanding of the original author's t, I visited Mao Shan in September 2014. I visited with the Taoist ster Feung Ke Zu. Master Feung is the current assistant abbot the Mao Shan Temple. I did not have the opportunity to see the ef abbot, for he was away on business while I was there.

I owe my appreciation to my student, Leslie Mitchell, for her sistance in the translation of this book. She helped me organize flow of the book and together, we maintained the correct mmar, without compromising the knowledge within the ipture. Furthermore, she helped to prepare the book for blication. For her help, I am grateful.

Introduction

Nei Dan 內丹 is a small unit of qi which is embodied through the guidance of the practitioner's mind. In the practice of Nei Dan 丹, the practitioner's body serves as a cauldron. The unit of qi circulates through the meridian system, directed by the practitioner's conscious guidance. The Jin is primordial qi which is the essence that circulates through the whole body. If it is not used for sex, it will boost the body's energy.

The principle of Nei Dan 內丹 Practice is to reverse the biological procession and aging of the body. There is a Taoist saying, "Follow one's desire, bring a birth 順生凡; Reverse one's desire, become an immortal 逆成仙". It indicates that processing Jin into Qi is the foundation of Nei Dan Practice.

Processing Qi to feed Shen is an even more difficult course for the practitioner. The book, Hwang Tin Nei Jen Jing 黃庭內景經, the Scenes of Internal Spirits, reveals the secrets of this practice. When the body's Jin is vigorous, the practitioner will sense muscles twitching and tapping. When the body's Qi is abundant, the practitioner will have visions of corporeal light. When the body's Shen is consolidated, the practitioner will be able to enter absolute emptiness.

At length, this practice will allow the Jin, Qi, and Shen to form Holy Fetus. This is also known as the Elixir, Nei Dan 內丹. This method of cultivation is called Nei Dan Gong 內丹功, Internal Elixir, Nei Dan Practice. This practice is divided into four stages: Building the Foundation 築基; Processing Jin to become Qi 煉精化氣; Processing Qi to feed Shen 煉炁化神; Processing Shen to be united with the Void 煉神還虛.

Building the Foundation

Since the Nei Dan practitioner's body serves as a cauldron. The fire is the practitioner's conscious mind. The fuel is the practitioner's breath. The practitioner's biological essence is cooked within the cauldron.

Physically, the heart is above the kidney. In Nei Dan Practice, the practitioner visualizes the kidney on top of the heart. 'San Shi Liu Ze 三屍六賊' is very harmful to the body. In order to build the Foundation, the goal of the practitioner is to kill 'San Shi Liu Ze'. 'San Shi', the three corpses, is the three deteriorated things – jin 精 qi 炁, shen 神. 'Liu Ze', the six thieves, or the six excesses - wind, cold, heat, damp, dry, fire. The practitioner must restore their body's condition. They do so by boosting their Jin, Qi, and Shen. This simply means that they take care of their body's health, laying a good foundation for Nei Dan Practice.

Middle or old aged people have worn out their qi and blood. These people must restore their health, then they can start a serious cultivation. In modern days, cancer patients can be regarded as carrying 'San Shi Liu Ze' in their body. Before they practice Nei Dan Gong, the cancer must be cured.

Mental training is of equal importance to the recovery of health. This is called Nian Ji 煉己, Curb One's Mind. Both mental and physical healing should progress equally. This is the process for Building the Foundation.

Processing Jin to Become Qi

The emission of the body's essence, or Jin, is called Lo 漏, leaking. When the body's essence remains in the body, it can boost the internal organs and the four limbs. If it leaves the body, the body must reproduce this essence. Constant emission of Jin is rmful to the health of the body. The body is unable to boost elf, because its energy must be used to reproduce the lost Jin.

Stopping the Lo 漏, or leaking is an essential task for the actitioner. Often, the practitioner is requested to shun sexual tivity for 100 days. The practitioner's mind focuses on the lower n Tien, which is the width of your four fingers below the bilicus. They focus their breathing in this area, moving the qi it up and down with the guidance of their mind. The actitioner will feel their lower Dan Tien warm. The theory of this actice is to use the breath in the lungs to press the diaphragm wnward. This causes the diaphragm to force the air down ong the internal organs.

Science tells us that matter can be in three states; gas, liquid, d solid. The air forced among the internal organs is in a gaseous te. It must be condensed to a liquid and condensed further to come a solid. The resulting tiny solid particle is called 'Shu Mi', millet, or 'Ming Zhu', the shining pearl. It can enter and pass ough the core of the spine.

Generally Xiao Zho Tian 小周天, the Small Heaven Circuit rts from the base of the practitioner's spine, ascending up the vernor Meridian 督脈. When the practitioner feels a thrusting sation in his tail bone, it is the movement of the solid particle of qi unit. Taoist call it 'Shu Mi 粟米', the millet, or 'Ming Zhu 明 , the shining pearl. It is also known as elixir, which results from practitioner continuing to refine his qi to be the smallest unit.

As it tries to pass through the 'Wei Lu Xue 尾閭穴', the tail ne, the practitioner uses his mind, consciously to guide this llet, which is a combination of Jin and Qi. Then the practitioner ll feel the thrusting sensation in the middle of the spine, 'Jia Ji e 夾脊穴'.

Next, the practitioner will feel the thrusting sensation in the base of the head, or the 'Yu Zhen Xue 玉枕穴'. Then the current of energy will move over the crown of the head and rain down through the eyes. It will enter the practitioner's throat and lungs reach to the middle of the chest at the middle Dan Tien. The shining pearl remains solid along its pathway from the 'Wei Lu Xue', the tail bone, until the time it reaches the eyes. At this point becomes a liquid flow, until it reaches the middle Dan Tien 中丹田 At the middle Dan Tien, the shining pearl, or Ming Zhu becomes fertilized egg nourished by its own liquid flow.

This entire process is called Tong Guan, Passing the Gates. The elixir descends along the front torso, through the Conception Meridian. It also passes through the three Dan Tien separately, at the Ni Wan, the brain; the Hwang Tin, the center of the chest; and the Ji Zhong 臍中, the umbilicus. Passing through the three gates is called Jin Yang Huo 進陽火, Send In the Yang Fire. Descending through the three Dan Tien is called Tui Yin Fu 退陰符, Retreating in Accordance of Yin. This is the terminology of Nei Dan Practice

Behind the eye lids, the practitioner will see a dim white light which shines three times. This is known as 'Yang Guan San Shien the three beams of the body's sun. These three beams shine white in color. Taoist call it 'Bai Xue 白雪', the white snow. It displays the energy of the lung, as white represents the lung. The practitioner has now completed the task of 'Nian Jin Hwa Qi', processing Jin become Qi. Now, the practitioner has completely recovered from blemishes in his health.

Processing Qi to Feed Shen

This stage of training focuses on feeding the qi to the shen so that they are one unit of jin, qi, and shen, in anticipation of producing the Holy Fetus. The secret training of the Nei Dan

...actice has now entered the Large Heaven Circulation. During ...e Small Heaven Circulation, the practitioner guides the elixir ...to orbit through the Governor and Conception Meridians. ...owever, during the Large Heaven Circulation the practitioner ...cuses his mind only on the middle Dan Tien, which will disperse ...e qi to the fourteen regular meridians and the six extra ordinary ...eridians.

The practitioner will see gold lights shine over his head. The ...ht will surround his whole body. In Taoist terminology it is ...lled 'Hwang Ya 黃芽', the yellow sprout. This is the display of ...leen qi, as yellow represents the spleen. The practitioner will ...so sense the whole body vibrating and the body will experience ...rious involuntary movements while in mediation. The book, ...wang Tin Nei Jen Jing 黃庭內景經 , *the Scenes of Internal Spirits*, ...scloses the secret of how the internal spirits are fed and how ...ey thrive through the practice of Large Heaven Circulation. This ...age is called 'Huai Tai 懷胎', Conceiving the Fetus.

The book, *Hwang Tin Nei Jen Jing* states, "Ni Wan Bai Jie Je Yo ...en 泥丸百節皆有神", the brain and the hundred joints of the ...dy hold their own spirits. Initially, all the spirits of the torso will ...y homage to the spirit of the spleen. Then the spirit of the spleen ...ll ascend to the court of Ni Wan, the brain. Nine spirits live ...ere. Madame Ni Wan and her two assistants are the three ...ading spirits. Six other spirits follow them. Chinese people know ...em as San Fen Liu Po 三魂六魄, the three souls and six ...bordinate souls.

The hundred spirits consolidate into one spirit of the spleen, ...coming the Holy Fetus. Then the spirit of the spleen confers its ...ergy to the spirit of Ni Wan, thus growing to become the Holy ...by. Next the practitioner will learn how to employ this Holy ...by. It will exit from the crown of the practitioner's head. Then the Holy Baby will return to the body of the practitioner. This routine training will continue for twenty years.

Process Shen be United with the Void

The void indicates the space outside of the practitioner's body. Before the practitioner sends the Holy Baby out, it remains inside the practitioner's physical body in order to cultivate itself. Now, it is time for the Baby to exit from the crown of the practitioner's head and enter the Void. When the practitioner's crown opens, he will hear a thundering sound and feel as if his brain is being cleaved by a huge ax. However, the practitioner should not be horrified. The practitioner will also smell a pleasantly fragrant odor emitting from his mouth and nose. This is the sign that the Holy Baby 聖嬰 is ready to leave the practitioner's body. The practitioner will then see lights shoot out from the pores of his skin, surrounding him with light.

When the Holy Baby leaves the practitioner's body, the practitioner must maintain full control of the Holy Baby. At first, the practitioner will let the Baby walk a few steps away from him; soon he will call the Baby back into his body. Day after day, the practitioner trains the Baby to go further, for longer periods of time, before he calls it back. When the Baby becomes stronger, the practitioner can send the Baby out one mile or a hundred miles over land or through the air. By using this method, the Baby will not get lost. The Baby may also face many tests from evil spirits. The Baby's safety depends on the practitioner's focused mind. There are four destinies the practitioner can choose from for his Holy Baby. They are: relocation, occupation, reincarnation, as well as continuation. Relocation occurs when the Baby leaves the practitioner's body to abide in Heaven. Occupation indicates that the Baby leaves the practitioner to find a young and healthy body which does not have a soul dominating it. The Baby will enter and reside in this body. Reincarnation is when the Baby enters the cycle of rebirth. The Baby will leave the practitioner and go into a fetus that will be born soon. Continuation means that the Baby will not leave the practitioner. Instead, it will stay inside the practitioner in order to restore his health so that he may continue to live.

Master Sun Bu Er in Meditation

The Authors and Interpreters

Madame Wei Hwa Tsun 魏華存
Author of *Hwang Tin Nei Jen Jing* 黃庭內景經

Madame Wei Hwa Tsun's shrine at Mao Shan

Generally speaking, the Taoist method of cultivating health can be classified into two major schools of practice: Chuan Zen School 全真教 (All Truth), which is Nei Dan training, and Zen Yi School 正一道教 (Correct One), which is the study of sorcery. The knowledge of Nei Dan training derives from the book of Tsan Tong Chi 參同契 (The File of Acknowledged Research) by Wei Bo Yang of the Han Dynasty. Its core training is: Nian Jin Hwa Qi (Refine the Jin to Be the Qi), Nian Qi Hwa Shen (Refine the Qi to Be the Spirit), Nian Shen Hwan Xu (Refine the Spirit to Return to the Void). Hwang Tin Nei Jen Jing shares only a part of its core training with Nei Dan practice. The author, Madame Wei Hwa Tsun, changed "Nian Qi Hwa Shen" (Refine the Qi to be the Spirit), to "Tio Qi Yang Shen 調氣養神" (Process the Qi to Feed Spirit). Therefore, in regards to the realization of immortality, the lessons taught in Madame Wei's school lead to a separate path than that of Chuan Zen School.

She also mastered Taoist sorcery which was similar to that taught at Zen Yi School. Madame Wei Hwa Tsun lived during the Jin Dynasty, around 265-318 AD. She was married and had three sons. Madame Wei was very fond of studying Taoism. One night, she was sent a dream from heaven, in which she was given the knowledge of becoming an immortal. She recorded the instructions from the dream. This record became Hwang Tin Nei Jen Jing.

Madame Wei was raised in a powerful and wealthy family. She gave up the power and wealth bringing her three sons to live a reclusive life in the Mao Shan. This is where she founded a Taoist school, which still exists and is in practice today. It is known as Mao Shan School (Reed Mountain School). Mao Shan is located in the county of Ju Rong 句容 of Jiang Su province 江蘇省.

Pu Ming Chan Shi 普明禪師
Author of *The Method of Mind Discipline* 煉心法

Pu Ming was a Buddhist monk of Chan Zhong school 禪宗. He lived during the Chin Dynasty. The exact time of his life is unknown.

Chan Zhong school's method of cultivation is a combination of Taoism and Buddhism. It was founded by Poti Dharma 菩提達摩 during the Nan Bei Dynasty of the Emperor Lian Wu Ti 梁武帝, approximately 500-580 AD.

This school became popular in the Tang Dynasty around 648-904 AD. Later, Chan Zhong 禪宗 was introduced to the Japanese. It was known as Zen in Japan, where it quickly became popular practice and belief.

Wu Chong Xu 伍沖虛
Author of *The Secret Phrases for the Three Levels of Nei Dan Training* 三乘秘密口訣

Master Wu was a Taoist of Chuan Zen School 全真道教. He lived during the late Ming Dynasty 明朝, around 1600-1685 AD.

He was a very famous and enthusiastic practitioner of Taoism. He spent a large part of his wealth seeking to learn the arts of Nei Dan School. Often he encountered inferior masters who bragged of their ability. Later, Wu Chong Xu found their knowledge deficient compared to his own. They cheated him of his wealth without imparting wisdom. This gave him deep sorrow.

However, he continued to pursue the truth of Nei Dan training. Eventually, he became enlightened with knowledge of the whole procedure. He dictated his thoughts to his disciple, Gu Yu Sau 顧與, who wrote it down.

Len Chan 冷謙
Interpreter of *Hwang Tin Nei Jen Jing* 黃庭內景經
Interpreter of *The Secret Phrases for the Three Levels of Nei Dan Training* 三乘秘密口訣

Len Chan was a Taoist of Zen Yi School 正一道教, which allows the practitioner to be a family man. He lived during the early Chin Dynasty 清朝, around 1600-1695 AD. He was a famous scholar of health studies and an expert in Qi Gong training.

Hwang Tin Nei Jen Jing 黃庭內景經
(Interior Yellow Court Scripture)
The Scenes of Internal Spirits

Shrine of Sun Chin, Yu Chin, and Tai Chin at Mao Shan

上清紫霞虛皇尊　太上大道玉辰君　閒居蕊珠作七言

s Majesty, the Superior Purity, Purple Sunset Emperor of
e Void of the Universe. He holds the supreme Tao, lives in
e court of Yu Chen Juen (jade temple constellation). In his
sure time, he composes poems of seven-word construction.

te: The original text of this manuscript was written in seven
inese character lines. This poetic format is called Chi Yen Shi 七
詩 (seven-word construction poem). Generally speaking, Taoism
ludes two parts: Tao Ja 道家 (Taoist masters), the theoretic

founders of the Tao; and Tao Jio 道教 (Tao religion), which implements the Tao. The Superior Purity is one of the Gods of Tao Religion. In Chinese, He is called Sun Chin 上清 (Superior Purity) and He is equivalent to Jehovah in the Judeo-Christian religions. In Taoism the trinity is Sun Chin (Superior Purity) ~ the Father, Yu Chin 玉清 (Jade Purity), or Yu Hwang Da Ti 玉皇大帝 (The Great Jade Emperor) ~ the Son, and Tai Chin 太清 (Extreme Purity) ~ the Holy Ghost. These three are united in one.

2. 散化五行變萬神 是為黃庭曰內篇 琴心三疊舞胎仙

Its name is the Interior Yellow Court Scripture, which provides insight into the ten thousand mysteries which evolve through the transformation of the five elements. When the practitioner's heart meets with the harmony of the lute's tone, the fairy will dance upon the three shelves of the uterus (Dan Tien).

Note: The five elements 五行 are symbols of the theory of absolute. They are represented by aspects as: 1. metal 金, wood 木, water 水, fire 火, soil 土; 2. white 白, green 青, black 黑, red 紅, yellow 黃; 3. lung 肺, liver 肝, kidney 腎, heart 心, spleen 脾. The uterus 胎仙 is the organ of reproduction which is like the Dan Tien's 丹田 generation of Da Yao (large medicine) & Xiao Yao (small medicine).

3. 九炁映明出宵間 神蓋童子生紫煙 是為玉書可精研

The extreme Yang qi emits light in the sky. The spiritual boy manipulating the process and emanating purple fog. All has been written in the Jade Book, so that the practitioner can carefully study it.

Note: In the original text Yang Qi is written as Jio Qi 九炁 (nine qi). Nine is the highest number of Yang 陽. Liu 六 is the highest

...umber of Yin 陰. The Jade Book is this book, Hwang Tin Nei Jen
...g (Interior Yellow Court Scripture).

詠之萬過昇三天 千災以消百病瘥 不憚虎狼之凶殘
以却老年永延

...y reciting this scripture ten thousand times, he will be lifted
...to the third level of heaven. Even if he is destined for a
...ousand disasters and a hundred illness in his lifetime, all
...ll be pardoned and erased. He has no need to fear of the
...uelty and ferocity of a tiger or wolf. His aging will slow
...own. He will easily reach longevity.

...te: This paragraph talks about the reason a Tao practitioner must ...udy this book.

上有靈魂下關元 左為少陽右太陰

...periority (upper) is the residence of the spiritual Fuen
...ul). In inferiority (lower), Guan Yuan (3 tsun below
...nbilicus) resides. Shao Yang (gall bladder) resides on the
...t. As Tai Yin (spleen) is at home on the right.

*...te: Fuen 魂 or Ling Fuen 靈魂 is similar to the soul in Judeo-
...ristian religions. According to this book, there are nine spirits in
...e brain. Madame Ni Wan 泥丸夫人 is in charge, followed by two
...sociates. Together, the three are called San Chi Ling 三奇靈 or San
...ng 三靈. All other spirits in the body are called Po 魄. Chinese
...ople commonly call them San Fen Liu Po 三魂六魄.*

后有密戶前生門

...Secret Chamber is located at the back. At the front, is the
...e Productive Door (genitalia).

Note: This paragraph introduces the position of the Hwang Ting 黃 庭 (yellow court, the middle Dan Tien). Mi Fu 密戶 (the secret chamber) is some time addressed as Yo Shi 幽室 or Yo Chue 幽關 in this book. Seng Men 生門 is the genitalia. Ming Men 命門 (the gate of life) is at the lower lumber area, about one inch above the sacrum.

7. 出日入月呼吸存

Whether it is one day or one month, on inhale or exhale, the focus must be on this place when practitioner does his perpetual breathing.

Master Tiger Skin Chang in Meditation

四炁所合列宿分　紫煙上下三素雲　灌漑五華植靈根

Four different qi combine to become one. Yet they are like stars in a constellation, each having their own station. Three unsullied clouds among purplish fog, irrigate the five flower plants. The roots of these plants have mystic power.

Note: Four different qi are metal (liver), wood (lung), water (kidney), and fire (heart). These surround the productive qi, soil (spleen). Three unsullied clouds are reference to the three Dan Tien. Five flowers indicate the five solid organs (liver, lung, kidney, heart, spleen).

七液洞流衝盧間　迴紫抱黃入丹田　幽室內明照陽門

From the orifices, fluid of seven elements flows out, surging into the cauldron. Spinning around the purple (heart) and embracing the yellow (spleen), they flow into the Dan Tien. In the quiet room, there is light which illuminates the Door of Yang.

Note: The orifices are glands. The fluid of seven elements 七液 is the heart, liver, spleen, lung, kidney, qi, blood. The Door of Yang 陽門 indicates the genitalia.

口為玉池太和官　漱咽靈液災不干　體生光華氣香蘭
滅百邪玉鍊顏

The mouth is a pond of jade. It is the office of great harmony. Drinking the mystic fluid, the practitioner will not suffer illness or disease. His body will glow with luster and emit a wonderful fragrance. A hundred evils will be vanquished accordingly. His face will shine like jade.

Note: Mystic fluid indicates the secretion which the practitioner produces upon completion the Xiao Zho Tian 小周天 training. His

internal organs generate secretions which combine to become a mil
粟米 *(elixir). It travels up from the base the spine, to the crown, down to the roof of the mouth, turning to liquid, which runs over t face. This is called mystic fluid. It is not salvia.*

11. 審能修之登廣寒　晝夜不寐乃成真　雷鳴電激神泯泯

By discerning and processing its value, he will be able to ascend to the palace Kuan Han (moon). In a trance, he will b baptized with lightning and thunder, without rest day and night. He is now True Man (immortal).

Note: Kuan Han 廣寒 *(moon* 月*): There is a legend about the beautiful Chan Er* 嫦娥, *who ate the mystic medicine. Her body became light and she ascended to the moon. The moon is cold and broad. The literal translation of Kuan Han* 廣寒, *in Chinese means broad and cold. It is a hint that the lifestyle of Taoist society is isolated and cold.*

12. 黃庭內人服錦衣　紫華飛群雲炁羅　丹青綠條翠靈柯

In the yellow court, there is a little man who wears colorful apparel. When he twists his skirt, a spinning cloud with a purplish luster manifests. He has a young figure like a lively growing green branch.

Note: The yellow court is the area of spleen. The spleen represents soil, its color is yellow.

13. 七蘥玉鑰閉兩扉　重扇金關密樞機

Then, he closes the door to the firewood storage, in which he keeps seven new born raccoons and a jade flute. Forcefully, I fans the golden gate which is a secret pivot.

te: The door to the firewood storage indicates the lower Dan Tien 丹田 which is 3 tsun below the umbilicus. The seven new born coons represent the seven senses: sight, hearing, speaking, smell, te, touch, intention. The jade flute indicates the genitalia. The den gate 金關 indicates the lower Dan Tien as well.

玄泉幽關高崔巍

e mystic fountain and the quiet door are against a high hill.

te: The lower Dan Tien's qi is the mystic fountain. The lower Dan n itself is the quiet door 幽關. After the completion of Xiao Zho n (small heaven circulation), the qi needs to pass through Wei Lu 閭 (tail bone). This seems like a high hill.

三田之中精炁微

the three Dan Tien, resides subtle jin (primordial qi) and qi elf.

te: Three Dan Tien indicate: the upper Dan Tien 上丹田, located ween the two eyebrows; the middle Dan Tien 中丹田 (the yellow rt), located between the two nipples; and the lower Dan Tien 下 田, located a hand's breadth below the umbilicus.

嬌女窈窕翳霄暉 重堂煥煥揚八威

girl with a slender figure is enjoying a sunbath. Soon, she plays her heroic manner in the grand hall.

te: The girl with a slender figure represents the blood 血. nerally, in scriptures about elixir refining, blood is represented by irl 姹女 or a tiger 虎. Qi is represented by a boy 嬰兒 or a dragon The spleen is the organ that generates the blood. The blood is ng boosted which is represented by the sunbath. This energizes the od in a heroic manner.

17. 天庭地關列斧釿 靈台盤固永不衰

A choice must be made between the court of heaven and the hall of hell. Axes and knives await the wrong reply. The correct reply brings a secure life on a platform which has a solid foundation and will never fall.

Note: Master Chang San Feng's 張三豐 *Wu Gen Su Hwa Chi* 無樹花詞 *(the poem about a flower tree without roots) says, "Shun Seng Fan* 順生凡*, Ni Chen Xian* 逆成仙*." which means "Following desire, an ordinary man produces children, not following desire, he becomes immortal". An ordinary man will go to hell* 地關*. An immortal man will go to heaven* 天庭*.*

18. 中池內人服赤珠 丹錦雲袍帶虎符 橫津三寸靈所居 隱芝翳鬱自相扶

In the middle pond there is a little man dressed in a colorful robe, wearing a scarlet pearl necklace, and a belt with a tiger head buckle. Under his San Tsun (tongue), full of saliva, the spirit lives. The hidden Ling Zhi (mushrooms) grow there, crowded and prosperous.

Note: The middle pond 中池 *represents the mouth. It is the same as the Jade Pond* 玉池*. The scarlet pearl necklace and the tiger head buckle represent the heart qi and blood respectively. The belt represents Dai Mai* 帶脈*, the girdle meridian. In the original text San Tsun* 三寸 *(three inches) indicates the tongue. It says that the tongue is connected with Dai Mai. The mushrooms are representative of the healing energy in this saliva.*

19. 天中之嶽精謹修 雲宅既清玉帝遊 通利道路無終休

The hill of heaven is the place the practitioner should endeavor to do his best. The house in the cloud has been kept

...at and tidy, the Jade Emperor wants to come for a visit. All ...ths are convenient without any obstruction, they are in a ...od state which will last forever.

...te: The hill of heaven indicates the core of the Dan Tien. The Jade ...nperor 玉帝, is known as Yu Chin 玉清 in this book.

. 眉號華蓋覆明珠　九幽日月洞空無　宅中有真常衣丹
　能見之無疾患　赤珠靈群華倩燦

...seems like an eyebrow above the shining pearl. In a quiet ...ve, the sun and moon hang inside. A True Man wearing a ...d robe lives there. His splendid skirt is studded with a red ...arl which makes him have a noble demeanor. If the ...actitioner can see him, illness and disease will occur no ...ore.

...te: The eyebrows indicate the lungs which are a little higher than ... heart. The shinning pearl indicates the heart. Anatomically, it ...pears like an eyebrow on top of heart. Quiet cave indicates the ...ple warmer. The sun and moon represent the heart and kidney ...pectively.

. 舌下玄膺生死岸　出青入玄二炁煥　子若遇之昇天漢

...neath the tongue and Xuan Yin there is a bridge that ...nnects the banks of life and death. A green color qi is issued ...th, a grey color qi is received in, these two qi are full of ...gor. If a scene like this is experienced, one will be able to ...cend into heaven.

...te: The practitioner must erect his tongue against the roof of the ...uth, it looks and acts like a bridge. Under the tongue 舌下 there ... two orifices which are called Jinjin 津津 and Yuyeh 玉液. Xuan ...1 玄膺 is the palate (the roof of the mouth). Green 青 qi is the

original qi among the internal organs. Grey 玄 qi is the qi of atmosphere.

22. 至道不息決成真 泥丸百節皆有神

The perfect Tao bestows believers with unceasing life, this is very true. Either in the brain or through a hundred joints of the body, all of these places have spirits which live there.

Note: The brain is called Ni Wan 泥丸 in this book.

23. 髮神蒼華字太元 腦神精根字泥丸 眼神明上字英玄
鼻神玉壟字靈堅 耳神空閒自幽田 舌神通命字正倫
齒神鍔鋒字羅千

The Hair Spirit, with thick hair, is named Tai Yuan (great leader). The Brain Spirit, with shrewd sense, has the name Ni Wan (pill of mud). The Eye Spirit, with upper brightness, is called Yin Xiang (flowery country). The Nose Spirit, with a jade platform, is named Ling Jien (assured information). The Ear Spirit, with empty spaces, is named Yo Tien (quiet farm). The Tongue Spirit, with the life channel, is known as Zhen Luen (correct ethic). The Teeth Spirit, with sharp and large tools, is named Lou Chien (thousands in order).

24. 一面之神宗泥丸 泥丸九真皆有房

It is the face which merges all spirits, and the spirit of the face pays homage to Ni Wan (the spirit of brain). In Ni Wan, nine spirits reside. Each spirit has its own chamber.

Note: Verse 81 of this text discusses Madame Ni Wan and eight persons dressed in white apparel. These are the nine spirits in the brain.

方圓一寸處此中　同服紫衣緋羅裳　但思一部壽無窮　非各別俱腦中　列位坐次向外方　所存在心自相當

The chamber in which one spirit resides is about one tsun (this is a Chinese measurement of approximately one inch) in diameter. All spirits wear red silk apparel and ponder the matter of striving for longevity. They do not live separately, but together in the brain. Seated in a line, they face outward. They undertake equivalent tasks, acting as the spirit of the heart.

Note: Ancient people believe that thought is a conjoined function of the heart and brain.

心神丹元字守靈　肺神皓華字虛成　肝神龍煙字含明
鬱道煙主濁清　腎神玄冥字育嬰　脾神長在字魂庭
神龍耀字威明

The Spirit of the Heart, holding a red elixir, is known as Sho Ing (inspiration guard). The Spirit of Lung, who holds a white flower, is named Xui Chen (empty achievement). The Spirit of the Liver is wrapped by dragon smoke and is called Un Ming (contain brightness). He is surrounded by the smoke of the Tao (purplish fog), which is responsible for separating the clean from the turbid. The Spirit of the Kidney, dressed in black, is known as Yu Yin (nourish infant). The Spirit of the Spleen, is named Fuen Tin (court of soul), he is on duty constantly. The Spirit of the Gall Bladder has a dragon's aurora and is known as Wei Ming (commanding presence).

六府五臟神體精　皆在胸內運天經　晝夜存之自長生

The Spirits of the Six Mansions and Five Warehouses have no bodily forms. Within each person's chest these Spirits operate the warp and weft of the heavens. Day and night, they undertake their task, in order that each person can enjoy longevity.

Notes: The Six Mansions are the six hollow organs: the stomach, large intestine, small intestine, urine bladder, gall bladder, triple-warmer. The Five Warehouses are the five solid organs: the heart, liver, spleen, lung, and kidney.

28. 肺部之宮似華蓋 下有童子坐玉闕

The palace of the Lung Spirit looks like a splendid canopy. Underneath it, a boy sits upon jade stairs.

29. 七元之子主調氣 外應中嶽鼻齊位

The nose is the son of the Lung Spirit. It is the Leader of the Seven and is in charge of breathing. It is the middle hill which merges with the Lung Spirit to perform the task.

Note: The Seven are the eyes, ears, mouth, nose, tongue, body and intention. Because the nose tip is ahead of them all, it is regarded as the Leader of the Seven.

30. 素錦衣裳黃雲帶

He wears a plain silk cloth which has yellow cloud patterns.

Note: This paragraph describes the appearance of the lung organ.

31. 喘息呼吸體不快 穩存白元和六氣

Panting and hasty breathing causes physical discomfort. The practitioner must focus on preserving the white leader (lung spirit) and the six qi (defense, blood, meridian, grain, lung, and ancestor).

Note: The six qi are: Wei 衛 (defense) qi, for defending against virus or impact injury; Xue 血 (blood) qi, for leading blood flow; Jinlou

eridian) qi, for repairing damaged health; Gu 穀 (grain) qi, for
pplying the body's energy; Hui 肺 (lung) qi, for breathing; and
ong 宗 (ancestor) qi, the original qi that a person receives from
eir parents.

. 神仙久視無災害 用之不已行不滯

dopt an everlasting introspection by following the practice
the immortals, and you will not incur harm or disaster.
ontinue this practice, and no stagnant qi will remain in your
dy.

. 心部之宮蓮含華 下有童子丹元家

tus flowers are waiting to blossom in the palace of the
art. Holding a red elixir, a boy stands underneath the
wers.

. 主適寒熱榮衛和

is in charge of the balance of cold and heat as well as
od and qi.

. 丹錦緋裳披玉羅 金鈴朱帶坐婆娑

white mantle covers a red shirt and skirt, in a posture of ease,
sits. Behind him is a gold bell girdled with a scarlet lace belt.

te: The white mantle indicates the lungs. The gold bell also
dicates the lungs, as the lungs are represented by metal. The
rlet lace belt is the heart. A large part of the heart covers the lung,
the belt girdles the bell.

. 調血理命身不枯 外應口舌吐五華

He helps to regulate blood and sustain life, thus the practitioner's body will not lose moisture. He reflects the glory of the five organ's essence through the mouth and tongue.

37. 臨絕呼之亦登蘇　久久行之飛太霞

When the life comes to an end, by just calling his name, a person's life will be resurrected. Continuing this practice perpetually, the practitioner is able to access the sky's extremity.

38. 肝部之功翠重裹　下有青童神公子

The palace of liver is surrounded by prosperous green, below that a noble young man stands clothed in green.

Note: The liver is represented by wood, its color is green. The noble young man is the Spirit of Liver.

39. 主諸關鏡聰明始

He is in charge of examining and reflecting on things. This is known as the initiation of intelligence.

40. 青錦披裳佩玉鈴

He wears green silk cloth and stands against large jade bells.

Note: In ancient China, only white jade was known of. Thus, the jade bells are white, representing the lung.

41. 和制魂魄津液平　外映眼目日月精

Pacifying the soul and the body's spirits, he also regulates the body's humor and blood. Outwardly, he shines forth throug

e essence of vision. The eyes may exhibit a luster like the
n and the moon.

. 百痾所鍾存無英　同用七日自充盈

hundred diseases are contracted due to the loss of liver
sence. To reverse it, cease to use the eyes for seven days, the
sence will become abundantly full, naturally.

*te: Taoists advise to listen inward and examine inward. Through
·ditation, one can restore lost liver essence.*

. 垂絕念神死復生　攝魂還魄永無傾

·surrection is possible by calling the spirit's name when life
bound to end. He helps to hold the Fuen (souls) and return
e Po (subordinate souls). A person's life will never end.

. 腎部之宮玄厥圓　中有童子冥上玄

e palace of kidney is a round building of mystery. A boy's
ure is within it, he is the Spirit of the Kidney. He is such a
ystic and unfathomable fellow.

*te: In the text, the boy is described as Xuan and Ming. Xuan 玄
ans mystery, Ming 冥 means darkness.*

. 主諸六府九液源　外應兩耳百液津

· dominates the Six Mansions and the source of nine fluids.
ternally, he is reflected in the two ears and the hundred
fices of liquid.

*te: The nine fluids are: tears, saliva, urine, stool, mucus, sperm,
eat, blood, and secretion.*

46. 蒼錦雲衣舞龍蟠

His ebony silk cloth surrounds him like a cloud, swirling like a dancing dragon.

47. 上致明霞日月煙　百病千災穩當存

Above, bright light from the sun and the moon shines through the fog like cloud. If the practitioner secures this well, he will be able to ward off a hundred illness and a thousand harms.

Note: This verse explains the importance of supporting the kidney and its essence. The sun light represents the heart and the moon light represents the kidney.

48. 兩部水王對生門　使人長生昇九天

Two kingly organs, belonging to the element of water, face the productive door. They both can help a person to preserve life in order to ascend to the ninth heaven.

Note: Two kingly organs indicate kidney and bladder.

49. 脾部之宮屬戊己　中有明童黃裳裡

The palace of spleen is associated to the element of soil. Inside, there is a brilliant boy who wears a yellow shirt and gown.

Note: The element of soil, in the original text, appears as Wu Ji 戊己, which means soil.

50. 消穀散氣攝齒牙　是為太倉兩明童　坐在金臺城九重

Consuming grain, dispersing qi, managing the strength of the teeth, two brilliant boys are the guards of Grand Granary. They sit on the gold platform at the center of the city which surrounded by nine walls.

Note: The two brilliant boys are the spirits of the spleen and the stomach. They are in charge of the Grand Granary. The city, is the spleen which is surrounded by all other organs. The spleen is related to soil, which is at the center of the other elements: water, fire, metal and wood. It is as if the spleen is surrounded by nine walls.

方圓一寸命門中　主調百穀五味香　辟卻虛羸無病傷

The center of spleen is a core, one tsun in diameter, called Ming Men. It is responsible for transforming a hundred grains into the five fragrant flavors. A weak and sickly person can become healthy and free of disease.

Note: Ming Men 命門, the gate of life is actually the back Dan Tien, at the second lumbar vertebra close to sacrum. In this verse, the spleen is called Ming Men 命門. This is only a nick name, which expresses its importance, not its proper location.

外應尺宅氣色芳　光華所生以表明

It refreshes the face with a new countenance, displaying the glorious luster of internal growth.

黃錦玉衣帶虎章　注念三老子輕翔　長生高仙遠死殃

He wears a yellow silk garment covered with a mantle of jade, his waist is girdled and secured with a tiger buckle. By concentrating and reciting the names of the three old men, the practitioner's body becomes light for flight, staying away from death.

Note: The tiger buckle indicates blood. The original text is Hu Zhang 虎章, which translates literally as "tiger seal." The three old men are the spirits of the heart, the kidney and the spleen.

54. 膽部之宮六府精　中有童子燿威明

The palace of the gall bladder holds the essence of the Six Mansions. Within it, there is a boy who demonstrates his bravery and wisdom.

Note: The Six Mansions indicate the six hollow organs.

55. 雷電八振揚玉旌

He produces lightning and thunder which causes the white banner to shiver and be blown to eight directions.

Note: This describes how a person's breath becomes agitated when they become angry.

56. 龍旂橫天擲火鈴

With a dragon banner crossing over the heavens, he throws out the fire bell.

Note: The fire bell indicates the heart. Anger causes the heart to expand and contract quickly.

57. 主諸氣力攝虎兵　外應眼瞳鼻柱間　腦髮相扶亦俱鮮

Dominating Strength waves the weapon to ward off the tiger soldier. He is displayed through the practitioner's eyes and pupils, even the nose, while anger arises to the brain which straightens the hair like fresh grass springing up.

Note: Dominating Strength is the gall bladder. The weapon is the heart. The tiger soldier indicates a viral invasion.

58. 九色錦衣綠華裙

He wears nine colors brocade shirt and a gorgeous green ski

te: His brocade shirt is filled with yang qi. This is indicated by *e* nine colors. Since the gall bladder is associated with the color *een*, the skirt represents the gall bladder. Physically, the liver is *ove* the gall bladder, so that the brocade shirt implies the liver.

. 佩金帶玉龍虎文 能存威明乘慶雲 役使萬神朝三元

rrying a gold bordered jade tablet, engraved with many *aracters* of dragons and tigers, he demonstrates his shrewd *artial* manner by riding a glorious cloud. He can manipulate *a* thousand spirits in the body to honor San Yuan, the three *ders*.

te: San Yuan 三元, *the three leaders, are sometimes called San* *u* 三老, *the three elders.*

. 脾長一尺掩太倉 中部老君治明堂 厥字靈元名混康
人百病消穀糧

e spleen is a foot long and covers the grand granary. In its *iter* is an old lord who manages the bright hall. His name is *in* Kang 混康 (general health), he is a sensitive one. He *ecializes* in healing hundreds of illnesses, and is good at *insuming* grains.

te: The grand granary indicates the stomach. The bright hall *plies* that spleen tends to be dry.

黃衣紫帶龍虎章 長精益命賴君王

earing yellow cloth and a purple belt, he has the demeanor *a* dragon or tiger. All of the tasks that generate body *ence* and lengthen life span depend completely on this *d*.

Note: The spleen is represented by soil which is the center of metal (lung), water (kidney), wood (liver), and fire (heart). Thus it is regarded as a lord.

62. 三呼我名神自通 三老同坐各有朋

'Call my name three times, unobstructed, all spirits can flow through the body'. Three old lords sit together, yet each one them has his own associate.

Note: The heart's associate is the small intestine. The kidney's associate is the bladder. The spleen's associate is the stomach.

63. 或精或胎別執方 桃孩合延生華芒 男女迴九有桃康

Either return to primordial qi or to conceive, each man has h choice. The fetus, seeming like a new peach, grows to form a splendid shape. Until at the ninth month, the coitus of a mar and a woman has a result, like a peach, the harvest is ready.

Note: The Taoist's have a saying; "Follow one's desire, bring a bir 順生凡; Reverse one's desire, become an immortal 逆成仙". The middle Dan Tien (yellow court) is regarded as the uterus for conceiving a spiritual fetus. Primordial qi is represented by Jin 精 (sperm) in the original text.

64. 道父道母對相望 師父師母丹元鄉 可用存思登虛空 殊途一會歸要終

The father (heart) and mother (kidney) of the Tao face each other, emulating my father and mother, the elixir is preserve in the country. In visualization, the baby ascends to the void All different paths lead to one place, now the task is comple

Note: The country indicates the upper Dan Tien (third eye), locate between the two eyebrows. The void indicates the heavens, outside the practitioner's body.

閉塞三關握固停　含漱金醴吞玉英　遂至不饑三蟲亡
意常和致欣昌　五嶽之雲氣彭亨　保灌玉盧以自償
色完堅無災殃

ocking three passes, keeping the hands held together, the
actitioner must dismiss all thought, then swallow the fluid
Jinjin and Yuyeh. Thus he will not suffer hunger and the
ree worms die. His mind and intention are in harmony and
s body becomes healthy. The cloud of five hills emit
undant qi, steadily it irrigates the jade house and the
actitioner's endeavor is rewarded. Five colors are in a
rfect state, disaster will stay away from him.

ote: The three passes indicate the mouth, the hands and the feet.
e mouth should not talk, the hands should be free of work, the feet
ould not move. Jinjin and Yuyeh, are the extra ordinary points
der the tongue. The three worms 三蟲 indicate cancer and related
seases which occur when the jin, qi, and shen deteriorate. Some
oist scriptures describe this as San Shi 三屍 (three corpses) and
u Ze 六賊 (six thieves). The three corpses represent the three
ferent cancer syndromes: Cancer Yin, Ju 疽; cancer Yang, Youn
; and the cancer which is neither Yin nor Yang. The six thieves are
 six excesses: wind, cold, damp, hot, dry, and summer heat. The
e hills indicate the five solid organs: heart, lung, kidney, spleen,
d liver. The jade house indicates the practitioner's body. The five
ors are the colors of five solid organs: the red of the heart, the
ite of the lung, the black of the kidney, the yellow of the spleen,
d the green of the liver.

上睹三元如連珠　落落明景照九隅　五靈夜燭煥八區
存內皇與我遊

serve the crown of the head, the three leader's qi can be
en as a strand of spinning beads. The bright aura shoots

through the nine orifices. The five spirits hold torches which illuminate the eight regions. The practitioner's mind rests in the interior yellow court and he is free to enjoy wandering with the interior emperor.

Note: Three leaders indicate the heart, the spleen, and the kidney. Nine orifices indicate the two eyes, the two nostrils, the two ears, the mouth, the urinary tract, and the rectum. Eight regions indicate the eyes and liver, the ears and kidney, the nose and lung, and the tongue and heart.

67. 身披鳳衣銜虎符 一至不久昇虛空

The practitioner's body wears phoenix apparel which expresses the tiger's demeanor. Once he has attained the Tao, he will be able to ascend to the void.

68. 方寸之中念深藏 不方不圓閉牖窗 三神還精老方壯
魂魄內守不爭競 神生腹中銜玉璫 靈注幽闕那得喪
琳條萬尋可蔭仗 三魂自寧帝書命

Random thoughts are concealed in the one tsun diameter cavity which it is neither square nor round, all windows and doors are shut. The three spirits have received the Jin (primordial qi), they become strong and mature. He who guards his inner souls, frees the souls from contest. The spirit in his chest are attractively dressed in jade. The practitioner's life will not be terminated, because his mind is focused on the quiet room. Ten thousand jade trees provide shade. The three souls find comfort. The Emperor will register the practitioner's name in the decree of fate.

Note: The quiet room indicates the upper Dan Tien (third eye), located between the two eyebrows.

靈臺鬱靄望黃野　三寸異室有上下　閒關榮衛高玄綏
房紫極靈門戶　是昔太上告我者

The mysterious platform is surrounded with prosperous clouds which face a yellow wilderness. In the space of three inches, there are separate upper and lower chambers. The leisure pass is high and glorious, which ensures ease and prospectiveness. The wedding room is an extreme purple color, which represents the most subtle gate. This, I was told by the Great Emperor Tai Chin in the past.

左神之子發神語　右有白元並立處　明堂金櫃玉房間
清真人當吾前　黃裳紫丹炁頻煩　借問何在兩眉間
俠日月列宿陳　七曜九元冠生門

The son of the left spirit (heart) proclaims mystic words, "At the right side, the white leader stands and leans side by side, there is a bright hall which has a jade chamber. A gold cabinet is stored in the chamber. When Sun Chin (Superior Purity) presents himself in front of me, wearing a yellow shirt and holding a purplish elixir, I feel his qi's perpetual vigor. May I ask who lives between the two eyebrows? He embraces the Sun and Moon and the constellations. He also commands Nine Leaders to produce seven shining auras. This surpasses what the productive door can do."

Note: This describes the Tao practitioner who has completed the Nei Tan 內丹 training. His holy baby lives in the bright hall, between the two eyebrows. The productive door indicates sexual reproduction. Because the Tao practitioner shuns sexual practice, he receives the great reward of being an immortal. Here, the Nine Leaders 九元 represent the nine spirits in the brain.

三關之中精氣深　九微之內幽且陰

Abundant Jin and Qi are deposited in the three passes. The nine cavities connect with the inner depth which is dark and deep.

Note: The three passes 三關 are the three Dan Tien 丹田. The nine cavities are the eyes, the ears, the nostrils, the mouth, the genitalia and the rectum. Though the orifices are external, their roots are connected to the internal organs, which are described as dark and deep.

72. 口為天關精神機　足為地關生命渠　手為人關把盛衰

The Heaven Pass is ruled by the mouth, which controls spirit essence. The Earth Pass is ruled by the feet, which is the channel of life. The Human Pass is ruled by the hands, which determine prosperity.

Note: This is a description of the practitioner's meditation posture. The tongue should be erect and touch the palate. The heels should be close to the root of the genitalia. The right hand is held with the thumb and index fingers forming a small circle; the left thumb penetrates the circle; with the left fingers covering the right fingers; the hands are laid in front of the lower abdomen.

73. 若得三宮存玄丹　太乙流珠安崑崙　重中樓閣十二環
自高至下皆真人　玉堂絳宇盡玄宮　璇璣玉衡色琅玕
瞻望童子坐盤桓

If the practitioner receives the mysterious elixir of the three palaces, the meteor will be like Taiyi 太乙 (Polaris), which is located in the constellation of Bei Dou 北斗 (the Big Dipper) lands securely on the top of the Mountain Kun Lun 崑崙山. The elixir flows down following the twelve winding stairs of the lofty tower. From bottom to top, there are spirits on guard. Both the jade hall (the lungs) and the scarlet house (the heart) nstruct the mysterious palace. It functions like the Xuan Ji 璣 (astronomy instrument), having the luster of jade, it flects the light. In sitting meditation the practitioner sees a y lingering around.

ote: *The meteor is the traveling elixir, which moves through back ne and flows down from face, transforming to liquid. The twelve nding stairs indicate the throat. The Mountain Kun Lun 崑崙山 a legendary Taoist mountain, here it represents the Middle Dan en (the yellow court). Taiyi 太乙 (Polaris) is the main star in the nstellation of Bei Dou 北斗, which means the north seven stars e Big Dipper).*

. 問誰家子在我身　此人何去入泥丸　千千百百自相連
一十十似重山

rious, the practitioner asks "Who is the son lives in my dy? How can he enter and reside in my brain?" A hundred a thousand things may be related with this fact. Ponder it counting from one to ten, the mystery seems to be hidden a series of mountains.

Taoist Master has conceived the Holy Fetus

75. 雲儀玉華俠耳門　赤帝黃老與我魂　三真扶胥共房津
五斗煥明是七元

A slender ear, a jade face, and cloud like hair unites as beauty. So the Scarlet Emperor, the Yellow Lord, and the soul joins together as one. In cooperation, these three immortals generate a house of saliva. In order for the five facial organs to reveal beauty, they must depend on seven things.

Note: The Scarlet Emperor is the heart. The Yellow Lord is the spleen. The soul is the liver because the liver stores the souls. The five facial organs are the eyes, the ears, the mouth, the nose, and the eyebrows. The seven things are sight, hearing, speaking, smell, taste, feeling, and touch. A healthy body and pure thought are reflected on the facial organs. In this way, a person can display his beauty.

76. 日月飛行六合間

As if soaring, the qi of the sun and moon travel among the six dimensions.

Note: The sun indicates the mind (intention). The moon indicates the breath (qi). In meditation, the practitioner's mind and breath travel among the internal organs (three dimensions) and time (past, present, and future: the other three dimensions).

77. 帝鄉天中地戶端　面部魂神皆相存

The Country of Emperor, the Center of Heaven, the House of Earth; all spirits of the face and even the soul derive sustenance from these three places.

Note: The Country of Emperor is the middle Dan Tien. The Center of Heaven is the upper Dan Tien, or the yellow court. The House of Earth is the lower Dan Tien.

呼吸元炁以求仙　仙公公子已可前

Through breathing the original qi to attain immortality, an elder immortal or a young immortal will present himself in front of the practitioner.

朱雀吐縮白石源　結晶育胞化生身　留胎止精可長生

Red Sparrow releases or withholds the essence of white rock, which is used to construct a new body, or to nurture a transformed body. By collecting Jin to preserve the holy fetus, one can achieve the goal of longevity.

Note: The Red Sparrow indicates the sexual organ and the white rock is the sperm. If the sperm is released, it is used to make a baby. If it is withheld, it is used to nurture the practitioner's body.

三氣右回九道明　正一含華乃充盈　遙望一心如羅星
室之下可不傾　延我白首返孩嬰

The three qi circulate in the right direction, the nine orifices emit brightness. Holding an unsullied thought, the practitioner is filled with Jin and Qi. With the mind focused on the internal organs, they are seen from a great distance, they are like constellations. Below the gold chamber, the secret should not collapse, it enables a white headed fellow to become a child.

Note: Qi circulation starts from the left and turns to the right, or in clockwise direction. The gold chamber indicates the three Dan Tien. The secret is the practitioner's focus on the three Dan Tien.

瓊室之中八素集　泥丸夫人當中立　長谷玄鄉繞郊邑

In the center of the jade chamber, Madame Ni Wan stands. She calls eight beings in white apparel to gather around her.

Some of them live in the long valley, some of them live in the remote country. Madame Ni Wan's command reaches to the farthest border.

Note: The jade chamber represents the brain. Please refer to verse 2 & 25 about the nine spirits in the brain.

82. 六龍散飛難分別　長生至慎房中隱　何為死作令神泣
忽之禍鄉三靈沒

The Six-Dragon Carriage is torn apart and scattered, it is hard to discern a difference. In seeking for longevity one should be extremely careful and restrain from the pleasure of the bedroom. Why must one cease life as the result and let all spirits cry? If this is ignored, it will bring disaster to country and cause three spirits decline and fall.

Note: The Six-Dragon Carriage represents the qi of the six Yin Meridians. This is practitioner's breath. If meditation breathing goes astray, the practitioner will not attain longevity, his life span will be no different than an ordinary person. The three spirits 三靈 indicate Madame Ni Wan and her two assistants.

83. 但當吸炁錄子精　寸田尺宅可治生

Simply absorb the qi, when you preserve your Jin. The one tsun space of the Dan Tien will enable a person to receive rejuvenation.

Note: The practitioner should be attentive and collect Jin 精 (primordial qi) at the time of Huo Zi Shi 活子時 (living 12:00 AM) This is the time of day when the genitals become energized.

84. 若當決海百瀆傾　葉去樹枯失青青　氣亡葉漏非己形
專閉御景乃長寧　保我泥丸三奇靈

a leak occurs, the flood can be compared to a sea wave. When leaves are lost, the tree soon withers, the green color can not be preserved. When liquid leaks, qi is lost, the bodily form does not look like the practitioner's own. Concentrate on sealing the leak and forsake fantasy, then the three precious spirits in Ni Wan (the brain) are preserved.

Note: *The three precious spirits* 三奇靈, *San Chi Ling is same as in verse 82, where the three spirits* 三靈, *San Ling are mentioned.*

恬澹閉視內自明 物物不干泰而平 確矣匪事老復丁
詠玉書入上清

With a placid mood, observing the inner self, naturally, the internal brightness can be seen. Free from the interference of external affairs, the mind will receive peace and harmony. When these instructions are implemented, the aged body will regain youth. Contemplating and reciting the Jade Book, the practitioner will be able to ascend and see Sun Chin.

Note: *Sun Chin* 上清 *means Superior Purity.*

常念三房相通達 洞得視見無內外

Constantly attend to the nourishment of three houses, they will have unobstructed connections. The view of the internal energy is as clear as the external.

Note: *The three houses* 三房 *indicate the heart, kidney and spleen.*

存漱五芽不饑渴 神華執中六丁謁

Preserve and swallow down the Jade Liquid to irrigate the Five Sprouts, neither hunger or thirst will arise. Shen Hwa (ring-bearer) will carry the robe, and Niu Ding (flower girl) will be the servant.

Note: The Five Sprouts 五芽, Wu Ya are the five solid organs. They are also represented as Wu Hwa 五華 (five flowers) in this book. Shen Hwa 神華 is a boy deity. Niu Ding 六丁 is a girl deity. Both of them appear to serve the practitioner's Holy Baby.

88. 穩守精室勿忘泄 閉而寶之可長活

Securely guard the chamber of sperm and do not let it be released randomly. Close the chamber and treasure it, longevity can be enjoyed.

89. 起自形中初不闊 三宮近在易隱括 虛無寂寂空中素 使形如是不當汙 九室正虛神明舍

The elixir is derived from the practitioner's corporeal form, it is the primitive essence which never escapes from the the body's control. The three houses are kept in a hidden place, but they can be easily accessed. The primitive essence comes from the void. Since it is clean, it must be allowed to remain unsullied. The nine chambers (the brain) are still in vacancy, they are a perfect fit for this Spirit (holy baby) to take as his lodgings.

90. 存思百念視節度 六府修治勿令故 行自翔翔入天路

A hundred memories and thoughts are preserved in the brain, yet they are used with limitation. The Six Mansions are managed and cared for very well, they are not allowed to return to their previous situation. Thus, the Holy Baby can ascend and fly along the path leading to heaven.

Note: The Six Mansions are the six hollow organs: gall bladder, small intestine, large intestine, stomach, urine bladder, triple-warmer.

治生之道了不煩　但修洞玄與玉篇　兼行形中八景神
十四真出自然　高拱無為魂魄安

The way of nourishing life is simple, not complicated. The practitioner should have knowledge of Taoism philosophy as well as this Jade Book. Furthermore, he pacifies the spirits of the eight directions. Twenty four immortals display themselves in the three levels of the Dan Tien naturally. Holding the hands in the greeting gesture and acting without expectation, the soul and spirits are at peace.

清靜神見與我言　安在紫房幃幕間　坐立室外三五玄
香執手玉華前　共入太室璇璣門

The god of peace and placidity presents himself, and speaks, rest your mind between the purple chamber and the curtain. Sitting down or standing up, meditate outside of the house for the duration of three to five incense burnings. Hand in hand we offer incense in front of the Jade Flower Tree. Then we walk into the grand chamber of the astronomy device together."

Note: The grand chamber of the astronomy device is the procession of Da Zho Tian 大周天, the universal functions of the human body.

高研恬澹道之園　內視密盼盡賭真　真人在己莫問鄰
處遠索求姻緣

Aiming at a remote goal, researching doctrine, I keep my mind placid, and roam in the garden of Tao. Earnestly, I have expectation of observing the genuine Fact. The True Man has come upon my own body, I don't need to ask of the matter from neighbors. Under this circumstance, should I still seek a marriage in a remote country?"

Note: Taoist art is devoted to the intertwining, or marriage, of the Kan 坎 (kidney, north, water) and the Li 離 (heart, south, fire) within the practitioner's body. This is not a corporeal union of male and female.

94. 隱景藏形與世殊 含氣養精口如朱

True Man lives in a hidden place and does not show himself. His behavior is certainly different from the people in the world. He is devoted to containing qi and boosting Jin. His lips are red, as if dyed.

95. 待執性命守虛無 名入上清死錄除 三神之樂由隱居 悠忽遨遊無遺憂

Fastening his Shin (intrinsic nature) and Ming (vital force) together, again, the practitioner must be void and empty of mind. His name has been registered onto the scroll of Sun Chin (Superior Purity). In the book of Dead Men, his name is not there. His three spirits enjoy the pleasure of a reclusive life. In a swift moment, he can enjoy an unworried time for traveling.

Note: Taoism doctrine focuses on the cultivation of Shin 性 (intrinsic nature) and Ming 命 (vital force). Simply speaking, a Taoist strives to be a good person with a healthy body. All other religions place emphasis only on being a good person. They don't place much emphasis on the health of the body.

96. 羽服一整八風驅 控駕三素乘晨霞 金輦正立從玉輿

Dressed in a feather costume, the True Man rides three white clouds, moved by the wind of the eight directions, to see the day break. Many gold carriages stand by, waiting to follow the Jade Chariot.

Note: This refers to how qi travels among the internal organs.

奚不登山誦我書　鬱鬱窈窕真人墟　入山何難故躊躇
間紛紛臭如奴　滄桑變幻同歸盡　此而不修胡為乎

Why not enter the mountain? You can have time to recite my book. This place has prosperous trees and plants fit for an immortal's lodgings. Why is it so difficult to make up your mind to enter the mountain? You are still hesitant to take the action. The affairs of the world are messy and uncertain. This is as intolerable as the smells from a slave's body. Land, with which the sea merges and mulberry farms, from which the sea retreats, all these changes will end. If you do not start to cultivate the Tao, what do you plan to engage in?"

五行相推返歸一　三五合氣九九節　可用隱地回八術

Escaping from the reciprocal dominance or creation of the five elements, you are devoted to merging with the True One. The qi of the Three Treasures and the Five Spirits are united. It meets with the lot of Ninety Nine. Then you can escape from your fate by hiding in an isolated land and living a reclusive life, then you can reverse the oracle of the Ba Gua 八卦 (Eight Hexagram)."

Note: The True One is the Tao. The lot of Ninety Nine, this is the lot of Chen (heaven, the extreme yang). Ba Gua 八卦 (Eight Hexagram) is evolved from Yin and Yang, which is the concept of creativity.

伏牛幽覺羅品列　三明出於生死際　洞房靈象鬥日月

Tame a bull in the quiet room, your name will certainly be included on the Immortal's List. Three bright lights emerge through the margin of life and death. In the Wedding Room, the Sun and Moon are guided to spin by the magnificent symbol."

Note: The bull is the practitioner's random thoughts during meditation. The quiet room indicates the lower Dan Tien. The three bright lights 三明 are a phenomenon that occurs when the practitioner has completed the practice of Xio Zho Tian 小周天. They are called Yang Guan San Xian 陽光三現 in the Taoist terminology. The Wedding Room 洞房 implies the middle Dan Tie the Yellow Court. The magnificent symbol is the Big Dipper. After the completion of 大周天 Da Zho Tian, the practitioner will feel as his torso is spinning. This is like the Big Dipper leading the Sun a Moon in circulation.

100. 父曰泥丸母唯一　三光煥照入子室　能存玄真萬事畢　一身精神不可失

"The father's name is Ni Wan 泥丸 (the brain), the mother's name is Wei Yi 唯一 (the one). Three bright things shine into the Room of the Son. If the True One can be preserved, you have already concluded ten thousand tasks which you anticipate to undertake. The Jin (primordial qi) or the Shen (original spirit) of your body should never be allowed to fall into a state of exhaustion."

Note: Wei Yi 唯一 indicates Tai Chi 太極. In the Tao Te Ching 道經, it says, "The Tao gives birth to the one 道生一." The three bright things are Jin 精, Qi 氣 and Shen 神. The Room of the Son the upper Dan Tien 上丹田 (the third eye), between the two eyebrows.

101. 高奔日月吾上道　鬱儀節璘善相保　乃見玉清虛無老　可以迴顏填血腦

he superior Tao, I know, is that the Sun runs to the Moon, d the Moon runs to the Sun at high speed. The practitioner ho preserves this art will receive a jade face countenance d a vigorous body. Those who can present themselves in ont of Yu Chin (Jade Emperor) will never become old. cause the Taoist's art can change a man's facial color and ost his brain with sufficient blood flow."

te: The Sun, Ze 日 indicates the heart. The Moon, Yue 月 dicates the kidney.

2. 口銜靈芒攜五皇　腰帶虎錄佩金襠　駕欻接生宴東門

our mouth secretes mysterious fluid which can revive the ve Nobles. Your belt is hung with a tiger-head buckle apped with gold. In a flash, you are transported to the clear st of day break, where you receive new life."

te: The Five Nobles are the five solid organs. The tiger-head ckle indicates blood circulation.

3. 玄元上一魂魄鍊　一之為物頗卒見　須得至真始顧盼 忌死氣諸穢賤　六神合集虛中宴

s mysterious Yuan Shen (original spirit) attains oneness, ur Fuen (soul) and Po (bodily spirits) are well refined. The eness can only be seen in quick glimpses. It requires your treme candor, then it grants your wish. Be cautious not to nge into dead qi, have a filthy body, or engage in mean ughts. The Six Spirits unite to enable the void to emit ht."

te: The Six Spirits are the eyes, nose, tongue, liver, heart, and 1g. The practitioner must use inner sight for introspection. The er sight focuses on the nose. The nose aims at the heart. rthermore, the eye represents the liver, the nose represents the lung, and the heart represents the tongue. *The practitioner straightens the tongue while in meditation.*

104. 結珠固精養神根　玉芪金鑰常完堅　閉口屈舌食胎津使我遂煉活飛仙

Being a conceiver and securing sperm, in order that I can nourish the root of Shen (original spirit). The Jade Vase and the Gold Key remain firm and in good shape always. I close my mouth and erect my tongue. I feed myself with the essence of the uterus (Dan Tien). Through cultivation, eventually I am able to become a lively, flying immortal.

Note: The Jade Vase and the Gold Key are the genitalia.

105. 仙人道士非有神　積精累炁以成真　黃童妙音難可聞玉書绛簡赤如丹　文曰真人巾金巾　負甲持符開七門

Neither Immortals nor Taoists have ever received special gifts from the gods. They accumulate sufficient Jin and Qi to become True Men. The mystic message of the Boy in the Yellow Court is difficult to comprehend. But the words in the Jade Book and the Scarlet Bamboo Tablet are written in cinnabar ink. The record says "The True Man puts a gold scarf on his head and wears armor. In his hand, he holds written charms. In this way, he opens the seven gates."

Note: The seven gates are sight, hearing, the spoken word, smell, taste, feeling, and sense. They are mentioned as "the seven things" in verse 75. It means that the True Man is very prudent in his thoughts and behavior.

106. 火兵符圖備靈關　前昂後卑高下陳　執劍百丈舞錦幡十絕盤空扇紛紜

order to attack the mystic pass, Fire Soldiers and charts are prepared. The front one is lofty and back one is below, it is obvious which one is high and which one is low. Holding a sword, the practitioner waves the brocade banner of a thousand feet. Ten sorts of desire are extinguished, emptiness soars up, the practitioner diligently fans his breath.

Note: *The Fire Soldiers 火兵 show that the practitioner maintains Wu Huo 武火, martial breath. The charts indicate the routes of the extra-meridians.*

7. 火鈴冠霄隊煙落 安在黃闕兩眉間 此非枝葉實是根

Fire bell's flame soars up and passes above the clouds, it soon turns to ash and falls endlessly. Landing between the two eyebrows, and on the Yellow Court. These two places are not for leaves or branches, they are places where the roots form when the practitioner engages in cultivation.

Note: *Fire Bell indicates the qi of the heart. The Yellow Court covers large part of the lungs, only the top of them can be seen. They look like two eyebrows.*

8. 紫清上皇大道君 太玄太和挾持端 化生萬物使我仙 昇十天駕玉輪

Emperor Sun Chin of the Purple Sunset is the great lord of Dao. He holds and balances Tai Xuan (great mystery) and Tai Ho (great harmony). These two can engender and propagate all existence, naturally enabling me to become an immortal. Ascending to the tenth heaven, I ride the chariot of the jade wheels.

109. 晝夜七日思勿眠 子能行之可長存

Through seven days and nights, the practitioner must remind himself not to enter sleep. He will be able to reserve an everlasting life, if he can carry out this session.

Note: This verse reveals the secret of collecting Da Yao 大藥 (large medicine) after the completion of Da Zho Tian 大周天 (Large Heaven Circulation).

110. 積功成煉非自然 是由精誠亦由專 內守堅固真之真 虛中恬澹自致神

Although it is not natural to accumulate qi and set it in procession by means of my piety as well as my concentration, I securely guard my inner self which is the true of truth. In the void I find placidity, of course I receive the marvelous gift.

111. 百穀之實土地精 五味外美邪魔腥 臭亂神明胎氣零 那從反老得還嬰

The kernel of a hundred grains is the essence of earth. All food provides five different flavors as well as external attraction. Actually, they will change, become foul and have bad odor. This will surely perplex the spirits and exhaust the qi of the uterus (Dan Tien). From where can I reverse aging, become an infant?

112. 三魂忽忽魄糜傾 何不食炁太和精 故能不死入黃寧

The three Fuen (souls) become sluggish and the bodily spirits become idle. Why doesn't the practitioner feed himself Yuan Qi which is the essence of Tai Ho (great harmony). Then he could be freed from death and his qi can enter and reside in the Yellow Court.

Note: Yuan Qi 元炁 (original qi) is the qi that is stored in triple-*armer.

3. 心典一炁五藏王　動靜念之道德行　清潔善炁自明光
起吾俱共棟樑　晝日曜景暮閉藏　通利華精調陰陽

*e heart manipulates the movement of qi and is the king of
*e five solid organs. Keep the qi in mind whether in action or
 rest, this is how to implement the Tao and its attribute Teh.
 one knows how to purify and keep good qi, his body will
*it light. Either standing up or sitting down, the heart qi
*ves with one's postures. It supports the body like pillars in
*ouse. It shines like the Sun in the day time. It retires for rest
* the evening. Therefore, it provides profuse Jin and balances
* and Yang in the body.

4. 經歷六合隱卯酉　兩腎之神主延壽　轉降適升藏初九
雄念雌可無老　知白見黑穩坐守

*ter experiencing the manifestation of the six extra
*ridians, I enter into retreat at the times of Mao (6:00 AM)
*d Yo (6:00 PM). The spirit of the two kidneys is responsible
* extending the span of life. A suitable amount of qi winding
* and down awakens the lurking Tsu Jio (initial nine). When
*ntemplating the male, I think about the female, I will not
*come old. When checking the white, thus I understand the
*ck. Certainly, I can preserve my game by waiting.

*te: This verse talks about the body sensation after the Du 督
*vernor) meridian and Ren 任 (conception) meridian become
*obstructed, the practitioner will experience the pulsing and
*ping sensations on the body. These sensations are aroused by the
*itement of the six extra ordinary meridians when they merge
*h the qi of the Du 督 and Ren 任 meridians. The six extra

meridians are: Yin Chao 陰蹻, *Yang Chao* 陽蹻, *Yin Wei* 陰維, *Yang Wei* 陽維, *Chong* 冲, *and Dai* 帶. *Retreat, in Taoist's meditation nomenclature is called Mu Yu (bathing). This means the practitioner can relax his mind to do breathing exercises. Male indicates the Yang Huo* 陽火 *(martial, male fire). Whereas, female represents the Yin Fu* 陰符 *(scribe, female disposition). White indicates the lung organ. Black represents the kidney organ.*

115. 肝之為氣調而長　羅列六府生三光　心精意專內不傾
上合三焦下玉漿　玄液雲行去臭香

The liver qi can be adjusted and extended. It is connected to the qi of the Six Mansions. When united well, the practitioner chest will beam with light three times. When the mind is focused on one intention, the qi of the internal organs will not be chaotic. Furthermore, its up-bearing is in harmony within the triple-warmer, and it joins with the down-bearing of Jade Syrup. This mystic fluid moves like a cloud and is able to remove the foul smell.

Note: Six Mansions indicate the six hollow internal organs. Light beaming three times is called Yang Guan San Xian 陽光三現 *(sunlight three beams). It is the sign of completion of Xiao Zho Tic* 小周天 *(small heaven circulation). Jade Syrup is previously mentioned in this book as jade fluid.*

116. 治盪髮齒煉五方　取津玄膺入明堂　下溉喉嚨通神明
坐視華蓋遊貴京　飄飄三帝席清涼　五色雲氣紛青蔥
閉目內視自相望

It cleans and heals the hair and teeth with the refined qi of the five organs combined. The Syrup is derived from Xuan Ying (the upper palate) and it is able to enter Ming Tang (the bright hall). Flowing down it irrigates the throat, making the mind

...ear and alert. Sitting under the splendid canopy, the practitioner can see it travel to and sojourn in the Gui Jing (the Metropolis). Three emperors take their seats and enjoy the cool breeze. The qi of the five-colored cloud rises like green lions springing up. When the practitioner closes his eyes, and turns his sight inward, he sees this scene happening within his chest.

Note: Ming Tang indicates the heart. Gui Jing reflects the Yellow Court. The three emperors are the heart, kidney and spleen.

7. 使心精神還相崇 七弦英華開命門 通利天道存玄根 二十年猶可還

The heart asks the other spirits for participation. They answer and give their assistance. The lively vibration of seven strings has opened the gate of life, which preserves the root of mystery. It provides unobstructed flow that connects the way of heaven. Even a person of one hundred twenty years of age can reverse his age to become a juvenile.

Note: The seven strings 七弦 are the five solid organs plus qi and blood.

8. 過此守道誠甚難 唯待九轉八瓊丹 要復精思存七元 日月之華救老殘 肝氣周流終無端

It would be rather difficult to preserve the Tao, if the practitioner ignores the Jade Book's guidance. He must go through the nine processes in order to produce an elixir with the luster of an eight faceted jade. Carefully concentrate to conserve the seven elements, because only the essence of the sun and moon can restore youth to an old and disabled man. The route of liver qi circulation has no beginning or end.

Note: The sun 日 *and moon* 月 *represent the heart and kidney. The seven elements are the five solid organs plus qi and blood.*

119. 肺之為氣三焦起　視聽幽冥候童子　調理五華精髮齒
三十六咽玉池裡　開通百脈血液始

The lung qi originates in the triple-warmer. The powers of vision and hearing await the direction of the boy in the quiet and dark room. It can regulate the essence of the five flowers and make the hair healthy and the teeth firm. The practitioner swallows the fluid of the Jade Pond thirty-six times, which provides humor and accommodates the flow of a hundred blood vessels.

120. 顏色生光金玉澤　齒堅髮黑不知白　存此真神勿落莫
當憶此宮有坐席　眾神會和輒相索　隱藏羽蓋看天舍
朝拜太陽東相呼　明神八威正辟邪

The boy's countenance shines as if made of gold or jade. It has firm teeth and dark hair. The practitioner must entertain this spirit and cannot be idle in serving it. He needs to know that there are many seats in the spirit's palace. Many other spirits visit and take their seats. The boy is concealed under the feather canopy and watches when the initial yang arises from the house of heaven. It bows to the grand yang of the east and shouts out loudly. It vanquishes evil affliction and becomes a brilliant spirit with commanding presence.

121. 脾神還歸是胃家　耽養靈柯不使枯　閉絕生門保皇都
萬年方祚壽有餘　是為脾建在中都

The stomach is the home where the spirit of the spleen must return. Take care of the chebule tree, lest it becomes dry. If the practitioner can close the door of reproduction, he can keep king's capital secure. A life span of ten-thousand years is possible, if the practitioner arranges his spleen properly in the central palace.

Note: The chebule (He Zi) tree has fruit which have the power to contract the lungs, restrain the intestines, and precipitate qi. In this text, the king's capital is the spleen.

2. 五臟六府神明王　上合天門入明堂　守雌存雄頂上光
方內圓神在中　通利血脈五藏豐　骨青筋赤髓如霜
救七竅去不祥

The five solid organs and the six hollow organs are the lords of the spirits. Their qi can trace up to Tian Men (the gate of heaven) and enter Ming Tang (the bright hall). If the practitioner knows how to guard the Female and preserve the Male, lights will shine from his crown. Externally, the Ming Tang is a box. Internally, it is a sphere. This is where the spirits dwell. They serve the blood vessels and boost the five organs, allowing the practitioner to keep his bone light green, his sinew pink, and his bone marrow frost-like. The spirit of the spleen can keep the seven channels from becoming exhausted and can ward off illness.

Note: Tian Men 天門 (the gate of heaven) is the fontanel at the crown of the head. Ming Tang 明堂 (the bright hall) is the Yellow Court. The Female indicates Wen Hou 文火 (soft breathing). The Male indicates Wu Hou 武火 (strong breathing). Healthy, living bone has a light green hue. Living sinew is pink and healthy bone marrow is frosty white.

The spirits of the five organs worship Madame Nei Wan

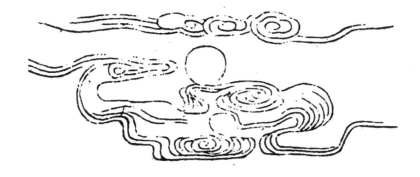

123. 日月列佈設陰陽　兩神相會化玉漿　淡然無味天人糧
子丹進饌肴正黃　乃曰琅膏及玉霜

The two qi are like the Sun and Moon representing the character of Yin and Yang. When the two spirits meet, their essences blend and become Jade Syrup. Although it is plain and tasteless, it is food for Heavenly Beings. The practitioner ingests his own elixir which is a dainty food with a yellow core. It is like a moist salve with the luster of jade or frost.

4. 太上隱環八素瓊 溉益七液腎受精 伏於太陰見我形
風三玄出始青 恍惚之間至清靈

hen the practitioner keeps extreme purity and hides the
[ja]de among eight white Chong Hwa 瓊花, the seven fluids
[be]come abundant and prevail. The kidney begins to deposit
[semen?]. When remaining in extreme darkness, one will be able to
[se]e the presence of his soul. Accompanied by a spring wind,
[th]e three mysteries emit new leaves. In trance, one comes to
[an] atmosphere which is so subtle and pure.

[No]te: Chong Hwa is recorded in historical books as a large and
[pr]etty flower which was grown in Yang Zho 楊州, a county on the
[ea]st coast of China along the Yangtze River. Here, Chong Hwa refers
[to] a person's white flesh. Three mysteries indicate Jin, Qi and Shen.

5. 坐於颷台見赤生 逸域熙真養華榮 內盼沉目煉五行
炁徘迴得神明

[Sit]ting on a platform, a flash of red is seen arising from the
[hu]rricane. In the land of leisure, prosperity is nourished by
[pe]aceful truth. When one lowers his sight and looks inwardly,
[one] can use spirit to refine the five elements. Three qi linger
[ar]ound the practitioner's corporeal form. He has received
[my]stic enlightenment.

[No]te: Here the five elements indicate the five organs: Lung
[re]presents metal, liver represents wood, kidney represents water,
[hea]rt represents fire, and spleen represents soil.

6. 隱龍遁芝雲琅英 可以充饑使萬靈 上蓋玄玄下虎章
浴盛節棄肥莘

[Th]e food for Heavenly Beings is like a hidden dragon, a
[el]usive Ling Zhi or a piece of marvelous jade. It can be

served to quench hunger and pacify ten-thousand spirits. Its influence ascends up to Xuan Xuan (Xuan Yin) and goes down to Hu Zang (the magnificent tiger). The practitioner must purify his body by bathing, forsaking meat, and abstaining from greasy foods.

Note: Xuan Xuan is Xuan Yin 玄膺 (the upper palate). Hu Zang 章 is the kidney.

127. 入室東向讀玉篇　約得萬遍義自解　散髮無慾以常存
五味皆至正炁還　夷心寂悶勿煩冤　過數已畢體神精
黃華玉女告子情　真人既至使六丁　既受隱芝大洞經

Entering the chamber, facing east, the practitioner reads the Jade Book. After reading it ten thousand times, he comprehends its meaning intuitively. He unties his hair and empties his mind of desire, so that he can preserve the Tao eternally. When he regains the correct qi, he will experience a pleasant taste of the five flavors. Soothing the mind and enduring seclusion, he will not suffer vexation and misery. After perusing the Jade Book many times, his body and spirit become refined. The jade girl of the yellow flower reports to him about his current internal condition. When he appears as the immortal, he will take command of Liu Ding. He has now received the Scriptures of the Large Cave. They are like food for the reclusive Ling Zhi.

Note: The jade girl of the yellow flower is young and beautiful. She represents the spirit of the spleen in this text. Liu Ding are the six servant girls which indicate the six hollow organs. Ling Zhi 靈芝 a type of healing mushroom. The Scriptures of the Large Cave 大洞經, indicate the scriptures of Taoism. They are classified in two parts: The metaphysics 洞玄 and the sorcery 洞神. Hwang Tin Ne Jen Jing pertains to the metaphysics 洞玄.

8. 十讀四拜朝太上　先謁大帝後北向　黃庭內經玉書暢
者曰師受者盟　雲錦鳳羅金鈕纏　以待割髮肌膚全
手登山歃液丹　金書玉簡乃可宣　傳得審授告三官
令七世受冥患　太上微言致神仙　不死之道此其文

The practitioner reads the Jade Book ten times, then bow four times to Tai Chin (extreme purity) and Sun Chin (superior purity). First, he pays a visit to the Great Emperor, Yu Chin (jade purity). To his teacher who bestowed upon him this book, he makes a vow. A robe of cloud brocade and phoenix silk, embroidered with gold thread at its border, is prepared for the immortal after his hair is trimmed and his body is purified. In ceremony, the great emperor holds his hand as he walks up the hill and drinks the liquid elixir. Now, it is time to disclose this Jade Book. When he has received this book, he discreetly passes it down to his own disciples. He must report his achievement to the three offices who are concerned. Otherwise, seven generations of his descendants will suffer the blame. The secret words of Tai Chin (extreme purity) and Sun Chin (superior purity) are only for those who are immortals and gods. The method of escaping death is included in this text.

Note: After attaining the Tao, the practitioner must report his achievement to the Chief of Taoists 掌教, the Emperor in Hell, Yen 閻羅, as well as Yu Chin 玉清, the Jade Emperor 玉皇大帝. There is a common Taoist saying about the Jade Book, "If you are not God, you are forbidden to read this book". This is understood to mean that if one has not attained the status of an immortal, the Secret Book is incomprehensible, even if one reads it. Therefore, only after one obtains immortality, will the time be right to disclose the Jade Book.

修心之秘

The Method of Mind Discipline

Verse 1
Before Herding

獰頭角恣咆哮　奔走溪山路轉遙
一片黑雲橫谷口　誰知步步犯佳苗

What kind of creature appears with a roar?
Its ferocious head has two horns.

It runs and walks along the trails of a creek or hillside.
Soon it becomes distant and remote.

It seems like a dark cloud,
The bull lays across the entrance of the valley.

Who would know what this creature is doing?
Every step of its walk damages the lively grain seedlings.

Verse 2
The Initial Discipline

有芒繩幕鼻穿　一迴奔競痛加鞭
來劣性難調制　猶得山童盡力牽

straw rope is tied to the harness.
e boy of the mountain penetrates the bull's nose
ith a ring.

e whip will land on its back,
the bull wants to have a run.

bad disposition is difficult to manage,
is happens from time to time.

ying his best to hold the rope,
e mountain boy must use his full strength.

Verse 3
Under Domination

調漸伏息奔馳　渡水穿雲步步隨
把芒繩無少緩　牧童終日自忘渡

Slowly the training goes, slowly the bull becomes obedient.
It stops charging and running.

Passing by creeks, going through clouds,
Step by step, it follows.

Still, the herding boy holds the straw rope,
He has no intention to loosen his hold.

All day long his mind is oblivious,
He forgets the matter of passing over the creek.

迴首第四
Verse 4
Turning Back Head

日久功深始轉頭　癲狂心力漸調柔
山童未肯全相許　猶把芒繩且繫留

Over a long period of time, the training
becomes effective,
The bull no longer wants to turn its head back.

Its excited and agitated mind calms down,
By and by it becomes peaceful.

Still the mountain boy cannot fully trust
and approve of its behavior.

He holds the rope in hand,
And ties it to a tree.

Verse 5
Obedient By Training

綠楊蔭下古溪邊　放去收來得自然
日暮碧雲芳草地　牧童歸去不須牽

Along the sides of an ancient creek, under the willow trees,
green shade is provided.

The bull is free to roam, and it can be called back,
it has reacted in an instinctive course.

Above the fragrant grass pasture, where they linger,
there are blue clouds and colorful sunset.

No longer is it necessary to hold the bull,
as the herding boy returns home.

無礙第六

Verse 6
No Inconvenience

露地安眠意自如　不勞鞭策永無拘
牧童穩坐青松下　一曲昇平樂有餘

Sleeping on the ground under an open sky,
The mountain boy enjoys himself, his mind is content.

Do not bother to use the whip,
The bull can now be kept without restriction.

Under the green pine,
The boy sits on the ground at ease.

A tune issues forth, bringing peace and cheer,
The pleasure is more than he has ever known.

運第七

Verse 7
Let the Fate Be

岸春波夕照中　淡烟芳草綠茸茸
餐渴飲隨時過　石上山童睡正濃

In spring, the light of sunset illuminates the bank of willow trees,
as well as the waves in the creek.

Smoke is seen rising in the distance,
vivid green grass is everywhere.

As time passes, quenching hunger and thirst,
the bull freely eats grass and drinks water.

Upon a large rock,
the mountain boy falls into a deep sleep.

相忘第八

Verse 8
Mutual Oblivion

白牛常在白雲中　人自無心牛亦同
月透白雲雲影白　白雲明月任西東

The white bull constantly roams
in white clouds.

Neither the boy's, nor the bull's
mind is attached to anything.

The moon beam shines through the white cloud,
the shadow of the cloud is white.

When the time to part comes,
the white cloud goes east, and the moon goes west.

Verse 9
The Solitary Moon Light

牛兒無處牧童閑　一片孤雲碧嶂間
拍手高歌明月下　歸來猶有一重關

Nowhere, can the bull be found, it has walked away,
At ease, the herding boy enjoys his freedom.

Among the green hills and valleys,
A lonely cloud soars and drifts.

Under the moon light,
The boy claps his hands and sings in a high voice.

Reflecting upon the sequence,
Still, they have neglected a gate (immortality) which must be passed through.

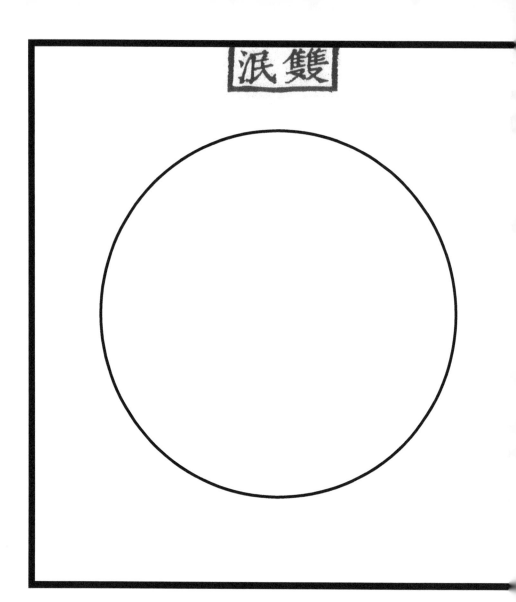

Verse 10
The Two Fade Into the Void

人牛不見杳無蹤　明月光含萬象空
若問其中端的意　野花芳草自叢叢

Both have faded, the boy and the bull,
No foot print is left on the ground.

The moon light shines upon all existence,
But they are nowhere to be found.

Some may ask about the real meaning,
The answer is in the waving sea of wild flowers and fragrant grass.

三乘秘密口訣並註

The Secret Phrases for the Three Levels of Nei Dan Training and Its Interpretation

*Jin Qi Shen:
Three Level Training*

吸自然神炁戀　陽生起火火方全　周天初用分子午
象陰陽六九連　約言百日是程期　精煉功勤化炁奇
炁居其須起脫　已成無漏欲遷移

...eathing with ease, without force, the practitioner's Shen 神 ...d Qi 炁 are in unison. The fire's 火 condition becomes ...rfect when the fire starts at the time of Yang 陽生 rising. ...th the commencement of Heaven Circulation 周天, they ...st practice at the time of Zhi 子時 and at the time of Wu 午... The lots 爻 of the Ba Gua 八卦 symbol are represented by ... 六 and Nine 九, repeating in a cycle. The diligent one ...ncentrates on the practice for approximately one hundred ...ys, they will then experience the wonder of transferring Jin

精 into Qi 炁. This is the Genuine Qi 真炁 which resides in th[e] umbilicus area (Dan Tien), and must be transcended. When the Lo 漏 is stopped and the practitioner becomes Leak Proo[f] 無漏, his focus must shift.

初乘小周天秘訣
The First Level

初乘小周天築基者 煉精化炁 閉關趺坐 於活子時清靜止念
垂簾塞兌 收視返聽 迴光於下丹田 以神馭氣而神入氣穴
以呼吸之炁而留戀神炁 方得神炁不離 升降自得 但炁有起止
起於虛危穴 坎宮子位 亦止於是 忌行有數 忌其太多 忌行有[?]
忌其太久

Those who practice the first level of Nei Dan 內丹 Training, Xiao Zho Tian 小周天, must focus on building the foundatio[n] 築基. The goal is to refine Jin until it becomes Qi. The practitioner must dissolve connections with external affairs and commence a sitting meditation. When the time of living Zhi Shi 活子時 (12:00 am) occurs, his mind must be tranquil and random thoughts must be suppressed. With eyes closed not hearing the outer world, his whole mind must be attenti[ve] to his inner self. The introspective lights illuminate the lowe[r] Dan Tien 下丹田. By using the Shen 神 to manipulate Yuan [Qi,] Shen is able to enter Qi Xue 炁穴, the cavity of Yuan Qi (belo[w] the umbilicus). Intentionally, the practitioner sends the qi of breath to merge with the Shen and Yuan Qi 元炁, thus Shen and Yuan Qi become inseparable. Either ascending or descending, the Shen and Yuan Qi enjoy a pleasant freedom[.] However Yuan Qi has points of initiation 始 and ending 止. [It] comes forth from Xui Wei 虛危 cavity (the pubis). This is the Kan Palace 坎宮, the reproductive door. After traveling

rough the body, Yuan Qi must return to its origination point
d stop there. The practitioner must limit manipulation of
ternal qi. Never practice too long or too many circulations.

單播弄後天炁者 恐以滯其先天之生機 後天炁用之不已
先天炁不旺 此修仙至緊至秘之功 故以周天三百六十限之
行三十六 積得陽爻一百八十數 午行二十四
得陰爻一百二十數 五位陽爻用九 故共一百八十數
卯時不同爻用 五位陰爻用六 故共一百二十數
酉時不同爻用 行沐浴以養之 古聖不傳火
云沐浴者不行火候也

ne should not indulge in using After Heaven Qi 後天炁, lest
hinder the productive motivation of Before Heaven Qi 先天
. When After Heaven Qi is continually set in motion, Before
aven Qi will not be prosperous. The Heaven Circulation
ould be confined to three hundred sixty breathing counts.
e practitioner can sum up the lot of Yang by the number
e hundred and eighty. At the time of Zhi Shi 子時 (12:00
), the practitioner must proceed with thirty six breaths.
haling to total the sum of the lot of Yang, one hundred and
hty. At the time of Wu Shi 午時 (12:00 pm), the practitioner
st proceed with twenty four breaths, accumulating the lot
Yin, which is one hundred twenty breathing counts. The lot
Yang, represented by Nine, which require five counts in one
ath. Therefore, there are one hundred eighty counts in
athing thirty six breaths. At the time of Mao 卯時 (6:00 am),
practitioner will not adhere to the idea of lots. The lot of
is represented by Six. One breath is five counts, thus the
m of counts in twenty four breaths is one hundred twenty.
e idea of lots is not used at the time of Yao 酉時 (6:00 pm)
her. The practitioner just proceeds with breathing as he

does Mu Yu 沐浴, bathing, in order to nourish the Dan Tien. The ancient sage would not pass (write down) the secret of Huo 火, fire. It is said that the practitioner does not apply the Huo Ho 火候 (fire and timing) during the time of Mu Yu, bathing.

行者積累動炁以完先天純陽真炁 凡一動一煉 積之百日則精不而返炁矣 百日築機 煉精化炁 乃大概言之 或有五六十日 或七八十日 得炁者 如年之衰老者 則二三百日 未可定也 功勤者易得 年少者易得

The practitioner perpetually activates the breathing qi, in order to initiate Before Heaven Qi, which is the pure Yang Qi 純陽炁. Once the Yang Qi is set in motion, one refines and accumulates it for one hundred days, then the Yang Qi of the body is fully revived. This one hundred days is used to build the foundation of vital force. This is an approximate time table for refining Jin into Qi. Some practitioners may only need 50 – 60 days. While others will need 70 – 80 days to obtain enough pure Yang Qi. If the practitioner is an aged person, they might even need 200 – 300 days. It is hard to make estimation. Naturally, it is easier to obtain sufficient Yang Qi if one practices diligently or if one is still young in age.

此時精已化炁 則無復有精 真炁已在臍之境矣 已得長生之基 為人仙也 故曰陽關一閉 個個長生 身已不死 而丹必可成也 是炁因靜定之久 不復動而化精 如有精則未及證於盡返炁也

At this stage the Jin has been converted into Qi, thus Jin will no longer be propagated. Genuine Qi has been stored in umbilicus area. The practitioner has completed building the foundation of longevity. He can now be addressed as the Immortal of Men 人仙. There is a saying, "Once the Gate of

...ng 陽關 has been closed, the practitioner enjoys longevity". ...nce his life will not end, he will definitely be able to attain ...e refinement of Nei Dan. However, if the practitioner ceases ...actice, and the Genuine Qi remains quiet and stable for too ...ng, it can transform, becoming Jin again. If the practitioner ...ll has Jin in his body, it is proof that he has not completed ...e work of refining Jin into Qi.

無漏者　則陰縮如小童子　絕無舉動為驗　便有止火之候

...ose who no longer leak Jin at all will notice that their ...nitals have shrunk to the size of a child's. His organ will no ...nger become excited. With this sign, it is time to move on to ...e termination of fire (stop breathing practice).

...時真炁亦不得死守於臍　須超脫過關名得金丹大藥
...以服食飛昇　故有三遷之法　既以七日口訣授天機　採其大藥
...五龍捧真之秘　度過三關以行中乘大周天之火候

...this time, the practitioner no longer needs to guard his ...bilicus attentively. He must transcend and go through the ...ree Gates and receive the Gold Elixir. This Great Medicine ...st be swallowed in order to ascend to heaven. There is also ...e secret of the method of three relocations. Here, I reveal ...aven's Secret of the Practice Phrases that will be used for ...ven days. The practitioner can proceed to collect the Great ...edicine by using this secret method which is called the Five ...agons Support Immortal. Thus, one can go through the ...ree Gates and enter the second level of practice, the Fire ...ning of Large Heaven Circulation.

三遷者 神在上田 炁在中田 精在下田 自下而遷中 自中而遷上 自上而遷出

The three relocations: The Jin, which resides in the lower Dan Tien, relocates itself to the Middle Dan Tien and becomes Qi. That Qi, resting in the Middle Dan Tien, then relocates to the Upper Dan Tien and becomes Shen. That Shen, resting in Upper Dan Tien, then moves out of the body and enters the Void.

七日口受天機 五龍捧真秘訣 秘密天機採藥收 蒲團七五火珠 三岔路上沖關妙 運轉真金神室留

The Secret of Heaven 天機 is tutored and practiced for seven days. This is the genuine secret of the Five Dragons Support Immortal. This secret of collecting medicine is the Secret of Heaven. One must sit on a Pu Tuen 蒲團, a grass meditation cushion, for thirty-five days, in order to experience the Huo Zu 火珠 (fire bead) traveling. The surging current against the Gate on the Three Meeting Way 三叉路 is so wonderful. The current transports the Genuine Gold 真金 to Shen Shi 神室, the mystic chamber, where it can dwell.

Note: Huo Zu 火珠 (fire bead), which is the Gold Elixir, represents the hot Qi current traveling in the practitioner's Du Meridian. Shen Shi 神室, the mystic chamber, represents the Middle Dan Tien.

七日是採大藥 七日之功 此萬古不洩之仙機 築基百日
查冥火炁薰蒸 故真炁因之忽然自有可見 故止後天炁之火
惟單採先天炁之藥 口訣採取於七日之內
因此時真炁盡歸於命根已腹內矣 雖有動 猶不離動處
只在內不馳於外

There is a schedule for collecting Great Medicine 大藥, practicing for seven days. This is the untold Motivation of immortals 仙機 throughout antiquity. The practitioner spends hundreds of days building a foundation. He will observe the fire and qi steaming and smoking in the periphery, then the Genuine Qi 真炁 is suddenly visible. At this point the practitioner can extinguish the Fire of the After Heaven Qi 後天炁. One only needs to collect the medicine from the Qi of Before Heaven 先天炁. This Secret Phrase is about collecting the medicine within seven days. Because the Genuine Qi has fully returned to the root of life after seven days training, the root of life is located between the umbilicus and lower abdomen. Though the motion of Genuine Qi can be felt in the lower abdomen, it will not move from this area. The motion is confined within the inner Dan Tien, it will not escape.

Note: *The root of life is actually located between the diaphragm and the center of the chest, the Middle Dan Tien.*

無火之火 無候之後 無火之火無候之後
言後天炁 後天之火當自陽光二現為始 至三現為終 故二現三現皆名止火之景 獨見陽光三現 方是採大藥之機 即用無火之火無候之候 此口訣中之口訣也

The practitioner should apply the Fire 火 of No Fire and the Timing 候 of No Timing. This means that he must now cease circulating the After Heaven Qi. The Fire of After Heaven initiates the Second Shine of Sun Light 陽光二現. The practitioner can end the fire (breathing) at the conclusion of the Third Shine of Sun Light 陽光三現. Thus, both the Second Shine and Third Shine indicate the time to terminate the fire. Only at the appearance of the Third Shine of Sun Light, does the practitioner have the opportunity to collect the Great Medicine 大藥. Therefore the phrase "Apply the Fire of No Fire, and the Timing of No Timing" reveals the most secret of the Secret Phrases 口訣.

Note: Huo 火, fire represents the practitioner's breath, whether breathing the atmospheric air, or breathing the Qi of the internal organs.

靜作蒲團取得下田先天真炁　名曰金丹　因採取之久　火候之足
還補炁之盛　謂之外丹　成其炁之發生　始有法成之妙相
而純陽之炁根始動　待到尾閭界也　在脊骨二十四椎至盡
三岔之路　有中左右三竅

Quietly sitting on Pu Tuen, the practitioner collects Before Heaven Qi from the Lower Dan Tien. This is called Jin Dan 金丹, the gold elixir. During the length of time it takes to complete the collection, its Huo Ho 火候, fire and timing matures. At that point, it is good enough to rejuvenate the practitioner's declining Qi. It is now called Wai Dan 外丹, external elixir. By utilizing the inception of Qi, the practitioner acquires a method to cultivate the elixir. When the Pure Yang Qi 純陽炁 is set into motion, it will linger around Wei Lu 尾閭, the tail bone area, as well as around the bottom of the 24th vertebrae. There at the three-Meeting path, appears three cavities. Each channel leads to either the center, the right or the left.

Note: The author indicates Wai Dan in this text. His meaning is different than most Taoist master's interpretation of Wai Dan. Most Taoists referring to Wai Dan are talking about the products of lead, mercury, rosin, alum, etc. By contrast the author here is referring the initial process of transferring Jin into Qi.

七日之功到五日之間 忽丹田如火珠 直馳上心即回
馳向外腎無竅可出 即轉馳向尾閭間沖關 此皆真炁自家妙用
由人力所至 但到關邊必用口授天機 方才過得關去
其真炁自然沖關 向上之機 加以五龍捧真之秘
龍即意土數五 蓋以意輕輕運動 則捧真陽大藥

When practicing the Seven Days Cultivation 七日之功, the practitioner must proceed until, on the fifth day, he feels the Dan Tien bring forth Huo Zhu 火珠, the fire bead, which flows upward toward the heart. Then it retreats, flowing down toward the genitals. There it cannot find an outlet. Thus it turns to surge against the Wei Lu 尾閭, the inner channel of the tail bone. At this moment, the practitioner must use the Secret Phrases 口訣 which were revealed to him, in order to address the Motivation of Heaven 天機. By doing this, the qi will be activated. Thus, the fire bead passes through the Gate. By causing the Genuine Qi to surge against the Gate, the secret technique of Five Dragons Support Immortal 五龍捧真 is applied. This means that the practitioner's mind is in association with the Spleen, the Intention of Earth 真意. Therefore, five is the number that should be used. The practitioner applies his intention and initiates a gentle lifting motion. Then he can bring up the Da Yao 大藥, Great Medicine, the Genuine Yang 真陽.

透尾閭夾脊玉枕三關 已通九竅 蓋每一關有中左右三竅
關則有九竅 直灌頂門 夾鼻牽牛過鵲橋 牛性主於鼻
牛之妄走 固有危險 故夾鼻使於當行之路
重樓乃喉之十二重樓 而入中丹田 神室之中
實以點化離陰 即乾坤交垢也

The practitioner's Genuine Qi has successfully passed through the Three Gates 三關: Wei Lu 尾閭, the tail bone; Jia Ji 夾脊, the middle spine; Yu Zhen 玉枕, the base of the head.

Actually his Qi has passed through a total of nine cavities, because each of these three Gates include three cavities. The Genuine Qi surges upward until it reaches the crown of the practitioner's head. Again it flows down and passes over the Chue Chao 雀橋, the magpie bridge. The nostrils must be contracted, as an ox with a ring through its nose is led over a bridge by its master. The Intrinsic Personality 元性 or demeanor of the oxen is connected to its nose. The ring prevents the ox from straying into danger. It is directed to a decent path. The Genuine Yang steps down Shi Er Chong Lo 十二重樓, the Twelve Stairs of the Tower. Then it enters the Zhong Dan Tien 中丹田, the Middle Dan Tien. This is the center of Shen Shi 神室, the Mystic Chamber. The Qi of Kan 坎, water, flows, indicating that the kidney has successfully transformed its nature. It has abandoned its nature of Yin. This is what has been called, 'Chen Kuen Jiao Go' 乾坤交姤, the Intertwining of Chen 乾, Heaven and Kuen 坤, Earth.

Note: Both Huo Zu 火珠, the fire bead, and Nu 牛, the ox represer the practitioner's Genuine Qi. Chue Chao 雀橋, the magpie bridg represents the upright tongue connected to the palate. In this position, the tongue seems like a bridge. Shi Er Chong Lou 十二重 樓, the Twelves Stairs of the Tower represents the practitioner's trachea.

以行大周天之火候 火原在下之物 合下田而行者也 雖合下而
時時充滿虛空 即有升降 而真我不動之元性 猶在於合下之內
古言心下腎上處 肝西肺左中 世人遂疑臍之上有一穴
如此則無根可歸 殆非也

Now, the training must focus on the procession of Huo Ho 候, Fire and the timing of Da Zho Tian 大周天, Large Heaver Circulation. The Huo or Fire is applied at the Lower Level.

atches with the Qi traveling in the Heaven of Lower Level. ough it matches with the Qi of the Lower Level, most of the ne the practitioner's breathing qi seems to enter the void d emptiness. Even as the practitioner manipulates his qi to cend and descend, the motionless Yuan Xin 元性, the imitive Intrinsic Personality is the True Self 真我. The True lf lives within the Lower Level (the Lower Dan Tien). The cient scriptures say, "The lodging of the True Self, is located neath the heart, above the kidney, to the west side is the er, the lungs are at the left and center". Thus the people of orld suspect that there is a cavity for meditation use, which ould be above the umbilicus. If this is true, then the True lf resides nowhere. The people of world know nothing at

乘大周天秘訣

e Middle Level

勤十月大周天 煉炁化神晝夜連 定力足時却世味
中遲速證胎仙

e Middle Level practice, is called Da Zho Tian (Large aven Circulation). It requires ten months diligent practice th intense concentration. Without an interlude, the actitioner refines the Qi to be Shen (spirit). The actitioner's mind is determined, day and night he practices thout rest. He will be able to shun the desire for food. How ll he does in his fast will indicate whether he will quickly slowly attain the immortality of uterus (Dan Tien).

中乘大周天火後者 煉炁化神 以周十日之天 用功無間
即古云功夫常不間 定息號靈胎 又曰晝夜晨昏看火候
不在吹噓併數息 蓋無間無時無數 為大周天之妙用
不似小周天之易行也 懷胎煉炁化神 入定者之候
其中三月定力而能不食世味 或四月五月或多月 始能不食者
功怠者得證果遲 惟絕食之證速 則得定 出定亦速 食為陰
有一分陰在則用一分食 由定而太和元炁充於中 則不饑不渴
若定心散亂 則有十月之外者 及不可計數 而使得定者
即歇氣多時 火冷丹力遲之故也

The Huo Ho 火候, Fire and the timing of the Middle Level D[a] Zho Tian 大周天 emphasizes the refining of Qi into Shen, in order to orbit the Heaven for ten months. The practitioner should practice diligently and unceasingly. An ancient sayin[g] states; "Continue the work without interlude, until your breath cannot be detected. Then you have achieved a Ling T[ai] 靈胎, "Uterus of Wonder". It is also said, "At noon, midnight[,] morning, and twilight, the practitioner devotes decent Fire [火] and Timing 候 in his practice". Then he can abandon blowin[g] (exhaling) and absorbing (inhaling) as well as counting. The wondrous thing that happens when the practitioner achieve[s] the Da Zho Tian is that he no longer needs to focus on timin[g] and counting, but can now let the qi naturally operate the course of Da Zho Tian. It is not like Xiao Zho Tian 小周天 which is easy to implement. After conceiving a fetus, the practitioner must focus to refine Qi into Shen, this is the opportune time for the practitioner enter Tranquility 入定. [He] must fast for three months. Some practitioners can even persist for four months, six months or even longer without food. He who is not diligent in the practice will be delayed i[n] the obtaining the Celestial Fruit 果證. He who dares to fast can obtain the Celestial Fruit sooner. Then, he can enter Tranquility. Also, he can retreat from Tranquility more

...ckly. Food has the nature of Yin. While the practitioner's ...dy still holds a portion of Yin, he is forced to take a portion ...ood. When the practitioner enters Tranquility, he can ...nsume the Yuan Qi 元炁, or Original Qi of Tai Ho 太和 ...eat harmony), which inflates the center of the corporeal ...dy. Then he would not suffer with thirst or hunger. If the ...ctitioner has a loose mind and cannot enter Tranquility, ...n he must practice even longer than ten months. The time ...ttain the Tranquility becomes remote. This is because he ...es interludes constantly. The Fire 火 and Elixir 丹 in his ...dy becomes cold and weak. However, he will eventually ...ain Tranquility.

...念除魔秘訣
...e Correct Thought For Distracting Evil ...sociations

...役景象屬陰魔　正念空空魔自瘥　呼吸無時神已定
...肖福長性靈和

...ere are ten thousand images pertaining to negative evil ...ociations. When the mind is empty, it can hold correct ...ught, and evil associations naturally retreat. When the ...ctitioner enters the state of breathlessness, his Shen ...eives complete tranquility. The evil associations dissipate, ...ss increases, the Intrinsic Personality attains spiritual ...mony.

...te: Shen 神 is the original spirit. Evil associations 魔 are random ...ughts and illusions.

正念除魔者 因神胎將完之時 外景頗多 有一分陰即有一分魔
或見其異而喜悅 貪見則着魔矣 見而不見 則不着魔 或聞奇異
或有可喜事物 或有可懼事物 或有可信事物 或有心生妄念
或有奉 上帝高真眾聖法旨 來試道行

When the Holy Baby is about to be birthed, the practitioner applies correct thought to ward-off evil associations, which are embodied in visions. If the practitioner still carries one portion of Yin in the body, he will be troubled with one portion of evil association. The practitioner will be delighted to see these strange things. If he craves seeing the images, he indulges the evil associations. If the practitioner is unconcerned with the presence of the visions, then he is free from the evil associations. These evil associations include; hearing unusual things, engaging in delightful matters, and experiencing horrid scenes. If the practitioner becomes convinced that these associations are real. He will manifest strange notions. He may imagine the he receives a decree from the Jade Emperor, or a notice from superior immortals holy men, that puts him to the test.

試過不着魔者 諸天保舉 或張妖邪來盜真炁 若心生一妄
穩提正念 眼見一魔 亦穩提正念掃去 靜中或見仙佛鬼神
樓台光彩 一切境界現前 一心不動 萬邪自退
只用正念煉炁化神 自然呼吸絕 而陰盡純陽 即無魔矣
然當魔過一次 則心愈靈一次 如得呼吸無 則炁不漏而返神
是真炁大藥服食已盡 炁已大定神全 煉炁化神之事始畢矣

After being put to the test, the practitioner who passes and is free from the evil associations 魔 will be recommended for attainment by the gods of Heaven 諸天. If the practitioner encounters the Yao Shie 妖邪 (the evil), who comes to steal his Genuine Qi 真炁, the practitioner's mind will develop random thoughts. He must quickly bring up correct thought 正念 to place those thoughts. In the quietness, he might see immortals, buddhas, ghosts, gods, or even splendid buildings. When any of these scenes are present, the practitioner must keep a steady mind to keep the ten thousand evil thoughts in retreat. Simply apply correct thought in order to refine the Qi to be Shen. When the natural breath has ceased, all of the Yin nature in the practitioner's body has been refined into Pure Yang 純陽. He is free from the evil associations. The experience of defeating the evil associations allows the practitioner's mind to be more spiritual and clear than before. If he attains the level of being breathless then his Qi becomes Leak Proof 不漏. He has completely converted his Qi into Shen 神. He can consume his own Genuine Qi. This is the Great Medicine 大藥. His Qi has gained great tranquility and his Shen is in a state of perfection. The process of refining Qi into Shen ends with this consummation.

乘度法超脫口訣
The Secret Phrases of the Upper Level Transcendental Method

月神全莫久留　由中遷上出重樓　依師度脫調神訣
載功成證果修

Once the perfection of Shen 神 has been completed, do not linger about for a long time. Using the route of the Twelve Stairs 十二重樓, the practitioner moves the Sacred Baby 聖嬰 from the middle Dan Tien to the upper Dan Tien. Now, the practitioner can move the Baby outside of his body. By following the mystic Secret Phrases 口訣 of the teacher, the method of guiding Yuan Shen 元神 is learned. It takes three years to complete the task. Then, the evidence of cultivation 證果 can be seen.

九年還虛口訣
The Secret Phrases of Returning to the Void

運用通神法妙圓　去留由己總隨緣　修成又有還虛理
面壁功深上界仙

By manipulating the method of connecting with the Almighty 通神, culmination has been attained. Either parting with the world or staying with it, this is up to the Yuan 緣, the Connection of Fate. In order to attain superior immortality, the lesson of Returning to the Void 還虛, must be completed. The practitioner will face the wall and train even longer.

上乘者神已純全　胎已滿足　必不可久留　如局於形中
而不超脫者　猶可離定而為動　則同於屍解之果而已　當用遷法

The practitioner who has completed the cultivation of the Upper Level Nei Dan 內丹, his Shen 神 is unsullied and perfect. His Sacred Baby 聖嬰 is well developed. In this circumstance, the practitioner should not keep the Baby in his body. If he does not let the Baby transcend 超脫, though it still moves and exercises, it is the same as the achievement of Shi Jieh 屍解, Freedom from Corpse. The practitioner must use the Methods of Relocation and Transcendence 昇遷之法.

以神之由中而遷於上田泥丸宮　既成純神　則謂之見性
加以三年乳哺　奶養神之喻也　當此上遷之時
非止拘神在軀殼之上　須用出神之理調神出竅　為身外之身
調神出竅亦須要知時　邱祖云若到天庭忽見天花亂墜始可出
是一至要之機

the practitioner applies his Shen to relocate the Baby from the middle Dan Tien to the upper Dan Tien, which is the palace of Ni Wan 泥丸宮. When the practitioner has refined the Pure Shen 純神, it is regarded as Jien Hsin 見性, Discovering Intrinsic Personality. Furthermore, he needs to Milk the Baby 哺 for three years. This is to nourish the Shen 神. When the practitioner moves the Baby to the upper Dan Tien, it is not meant to restrict the Baby's activity to the upper body only. The practitioner is applying the principle of Guiding the Baby out of the Chao 竅 (the Crown), through the upper Dan Tien. This allows the Baby to become the practitioner's other body. When guiding the Baby out of the Chao, it is important to determine the proper timing. Master Chiu Tsu Ji 邱處機 says, "When the Baby arrives at the Tian Tin 天庭, the Heavenly Court, the practitioner suddenly sees the Heavenly Flowers 花 falling down profusely in abundance. Then the practitioner can let his Baby fully show". This should be regarded as a very important note.

大危險之際初 謂其出而即入 不令出久 一步而即入
步而即入 亦不令見聞於遠境 調之久 出可漸久而復入
可漸見 聞於遠近而後入 不調者恐驟出外馳 迷失本性
於老成 必三年而後可

Although the Baby has been granted the ability to move outside of the practitioner's body, upon encountering great danger 大危險, it must be called back in immediately. It cannot be allowed to stay outside too long. The Baby must be called in after taking one walk 一步 or two walks 二步. Furthermore, It should not be allowed to listen from afar. If the Baby has been trained long enough, It can stay outside longer. However, It still needs to be called in. Even if the Baby is allowed to see and listen further away, still, It must be called back. If the practitioner ignores the training, when the Baby charges out suddenly, It may lose Its Intrinsic Personality 本性. However, it can take three years or longer for the Baby to build Its Personality to that of a worldly adult 老成.

凡初出者必調 依師度法出神 自上田出 念於身外
自身外收念於上田 一出一收 漸出漸熟 漸補漸足
如是謂知乳哺 三年而神圓 可以千變萬化 達天通地 報國濟世
超昇祖先 可舉念者 無不是神通妙用 欲少留則且止而佐時
欲昇驤則凌霄而輕舉 謂之神仙

The practitioner who starts to send out his Baby must have matured in his training. The practitioner receives his teacher instruction, the Transcending Method 度法, which states that he must send the Baby out from the upper Dan Tien 上丹田. The practitioner's mind moves the Baby outside of his body. Then he draws it back in and keeps it in the upper Dan Tien. Send out one time, draw back one time. Slowly the Baby progresses, gradually It becomes a worldly adult. Slowly the process becomes fortified, slowly it is strengthened. This is called Ru Pu 乳哺, Milking. It takes three years. The practitioner's Shen 神, or Holy Baby has the ability to perform one thousand or ten thousand transformations. Its power can extend world-wide, it can even reach up to Heaven. It can contribute Its power to help the country or to benefit the whole world. It can transcend the ancestors' spirits. When It has a thought, surely, it is a wonderful notion. For the benefit of the world, the Baby may want to live in the world for a period of time. When It wants to leave the world, It can easily ascend above the clouds. It is as entitled as Shen Xian 神仙, the mystic immortal, a god.

The Holy Baby ascends to Heaven

不欲住世 可用面壁之理 九年大定 煉神而還虛 可與上成仙佛
肩矣

he Baby does not want to reside in the world, the
actitioner can proceed to Mien Bi 面壁, Face Wall Sitting, for
ne years. He will receive great tranquility 大定. He has
ined his Shen 神, Holy Baby, and attained Huan Xui 還虛,
turning Void. He can now be regarded as Buddhas or
perior Immortals.

有真有入關之囑 故再及之 凡修行勿令人知 不近往來之衝
必遠樹林 絕其鳥風之聒 丹屋明暗適宜 牆心重垣 床坐厚褥
加以精潔芽菜淡飯 持素戒葷

There is a command that immortals have, which is stated as, Zu Guan 入關, Enter Gate. It is necessary to address this further. The practitioner should not let the world know of hi practice. He should avoid living near busy streets, and stay away from the forest, which has noisy birds and wind. His meditation chamber should have decent illumination, neithe too bright nor too dark. Heavy walls and a heavy bed cushic are desirable. He should not eat meat. Vegetable meals with clear bean sprouts, greens and rice are recommended for practitioner

小周天用功畢 清靜內守 謹言語 止諸事 行大周天時
宜同志三人互相守護 傾危須叫 冲虛伍真人曰 盡傳秘訣
以遇有緣者 因果必不昧也

When the practitioner has completed the lessons of Xiao Zho Tian 小周天, he needs to guard his inner self by living in a peaceful and harmonious environment. He must be careful about language and behavior. When practicing the lessons o Da Zho Tian 大周天, he needs to invite three comrades to liv with him. They will protect each other from dangers. Immortal Wu Chong Xui 伍冲虛 says, "All secrets have beer given to those who have the Connection of Fate 有緣 with m The relationship of Cause 因 and Fact 果 will not always be kept obscure 不昧."

大明崇禎七年歲次甲戌 上元門人 顧與發 恭錄

Respectfully recorded by Gu Yu Tau, the disciple of Immorta Wu Chong Xui in the Great Ming Dynasty during the seven year, or the Jia Wu year, of Emperor Chong Zen (1635 AD).

Index

ter Heaven Qi	87, 91
Gua	48, 85
Xue	6
ore Heaven Qi	87, 88, 92
Dou	39
Dipper	39, 49
use	104
estial Fruit	96
an Er	20
ang San Feng	1, 22
an Zhong	12
bule Tree	57
en Kuen Jiao Go	94
ef of Taoists	62
n Dynasty	12, 13
u Tsu Ji	101
Yen Shi	15
ng Hwa	59
ng Meridian	54
an Zen School	11, 12
e Chao	94
ception Meridian	6, 7
nection of Fate	100, 104

D

Dai Mai	22
Dan Tien	6, 7, 16, 18, 19, 21, 23, 31, 34, 36, 39, 40, 42, 45, 48, 50, 52, 86, 88, 90, 91, 93, 94, 95, 99, 101, 102
Da Yao	16, 52, 93
Door of Yang	19
Du Meridian	90

E

Emperor in Hell	61
Evil Associations	97, 98, 99

F

Fact	45, 104
Fire Soldiers	51
Five Dragons Support Immortal	89, 90, 93
Five Elements	16, 47, 59
Five Sprouts	43, 44
Five Warehouses	25, 26
Fluid of Seven Elements	19
Fuen	17, 29, 49, 52
Fuen Tin	25

G

Gate of Yang	88, 89
Genuine Qi	86, 88, 89, 91, 93, 94, 98, 99
Gold Elixir	89, 90, 92

Golden Gate	20, 21	Jade Vase	
Gold Key	50	Jia Ji Xue	
Governor Meridian	5	Jiang Su	
Grand Granary	30, 31, 33	Jia Wu Year	1
Great Danger	101	Jien Hsin	1
Great Medicine	89, 90, 91, 93, 99	Jin	3, 4, 5, 6, 11, 21, 34, 35, 36,
Guan Yuan	17		41, 42, 46, 48, 50, 53, 59,
Gui Jing	55		86, 88, 89, 90,
Gu Yu Tau	104	Jin Dan	
		Jinjin	23,

H

		Jin Yang Huo	
Han Ming	25	Jio Qi	
Huo Ho	88, 92, 94, 96	Ji Zhong	
Huo Zi Shi	42	Ju	
Huo Zu	90, 94	Ju Rong	
Huai Tai	7		
Holy Baby	7, 8, 37, 44, 98, 102, 103		

K

Hu Zang	60	Kan	46, 86,
Huan Xui	103	Kuan Han	
Hwang Tin Nei Jen Jing	1, 3, 7, 11,		
	12, 13, 15, 17, 60		

L

Hwang Ya	7	Large Heaven Circulation	7, 52,
			94,

I

		Leak Proof	86,
Immortal of Men	88	Li	
Intention of Earth	93	Lian Wu Ti	
Intrinsic Personality	94, 95, 97,	Ling	17, 22, 24, 25, 44, 59, 60,
	101, 102	Ling Fuen	
		Ling Jien	

J

		Ling Tai	
Jade Book	16, 17, 43, 45, 50, 55, 60, 61	Ling Zhi	22, 59
Jade Emperor	16, 23, 49, 61, 98	Liu Ding	
Jade Pond	22, 56	Lo	4, 5

ng Valley	42	Nian Shen Hwan Xu	11
u Chien	24	Nian Qi Hwa Shen	11
wer Dan Tien	5, 21, 40, 48, 86, 90, 92, 95	Nine Chambers	44
		Nine Leaders	37
		Ni Wan	6, 7, 24, 43, 48, 101
dame Ni Wan	7, 17, 24, 41, 42	Ni Wan Bai Jie Je Yo Shen	7
dame Wei Hwa Tsun	11, 12	Nu	94

O

o	53, 87
o Shan	1, 11, 12, 15
taphysics	60
thod of Mind Discipline	12, 63-83
thods of Relocation	
d Transcendence	100
en Bi	103
Fu	18
let	5, 20

Obscure	104

P

Po	17, 29, 49
Poti Dharma	12
Pu Ming Chan Shi	12
Pure Yang Qi	88, 92
Pu Tuen	90, 92

ng	29, 46
ng Dynasty	13, 104
ng Men	18, 31
ng Tang	54, 55, 57
ng Zhu	5, 6
tivation of Immortals	90-91
untain Kun Lun	38, 39
Yu	54, 88

Q

Qi	3-7, 11, 19, 21-24, 26, 27, 30, 34-38, 40-43, 46-60, 85-96, 98, 99
Qi Xue	86
Quiet Door	21

R

Red Sparrow	41
Ren Meridian	53
Ru Pu	102

n Bei Dynasty	12
Dan	1, 3, 4, 6, 11, 13, 37, 85, 86, 89, 100
Dan Gong	3, 4
n Ji	4, 6
n Jin Hwa Qi	6, 11

S

San Chi Ling	17, 43
San Fen Liu Po	7, 17
San Ling	17, 43

San Shi Liu Ze	4	Tao Jio	
San Tsun	22	Tao Te Ching	1,
Scarlet Emperor	40	Third Shine of Sun Light	
Scriptures of the Large Cave	60	Three Dan Tien	6, 19, 21, 38,
Second Shine of Sun Light	91	Three Fuen	
Secret Phrases	85, 91, 93, 99, 100	Three Houses	43,
Seng Men	17	Three Meeting Way	
Seven Strings	55	Three Old Lords	
Shen	3, 4, 6-8, 11, 35, 48-50, 59, 85, 86, 90, 95-97, 99-103	Three Passes	35,
		Three Worms	
Shen Hwa	43, 44	Tian Men	
Shen Xian	102	Tian Tin	
Shi Er Chong Lou	94	Time of Wu	85,
Shi Jieh	100	Time of Zhi	85,
Shin	46	Triple Warmer	
Sho Ling	25	True Self	
Shu Mi	5	Tsan Tong Chi	
Six-Dragon Carriage	42	Tsu Jio	
Six Mansions	25, 26, 29, 32, 44, 54	Tui Yin Fu	

W

Six Qi	26		
Small Heaven Circulation	7, 21, 54	Wai Dan	
Sorcery	11, 12, 60	Wei Lu	21, 92,
Sun Chin	15, 16, 37, 43, 46, 51, 61	Wei Lu Xue	
Superior Purity	15, 16, 37, 43, 46, 61	Wei Ming	
		Wen Hou	

T

Tai Chin	15, 16, 37, 61	White Leader	26,
Tai Ho	51, 52, 97	White Rock	
Tai Xuan	51	Worldly Adult	
Taiyi	38, 39	Wu Chong Xu	13,
Tai Yuan	24	Wu Huo	
Tao Ja	15	Wu Hwa	

...i Ji	30	Yu Chin	15, 16, 23, 49, 61
		Yu Hwang Da Ti	16
		Yu Yin	25
...o Zho Tian	5, 19, 21, 54, 86, 96, 104	Yu Zhen	93
...an Ji	39	Yu Zhen Xue	6
...an Xuan	60		
...an Yin	23, 60		
...an Ying	54	Zen Yi School	11, 12, 13
...i Chen	25	Zhen Luen	24
...i Wei	86	Zu Guan	104

Z

...ig Chao	54
...ig Guan San Xian	48, 54
...ig Huo	54
...ig Qi	16, 33, 88
...ig Wei	54
...ig Zho	59
...ow Lord	40
	17, 35, 42, 47, 53, 58, 87, 98, 99
...Chao	54
...Fu	54
...Wei	54
...Xiang	24
	53
...Chue	18
...Shi	18
...Tien	24
...n	35
...n Qi	52, 53, 86, 87, 97
...n Shen	49, 99
...n Xin	95
...Chen Juen	15